U0171191

微风吹拂

自在手账

[英]海伦·埃克斯利——编选
周成刚——主编

新星出版社 NEW STAR PRESS

Every day is a new day!

每一天，都是新的开始！

Mon. Tue. Wed. Thur. Fri. Sat. Sun.

Date

年

月

日

Live in the present, do all the things that need to be done. Do all the good you can each day. The future will unfold.

活在当下，为所应为。每日行善，未来会展现美好图景。

——和平朝圣者

（PEACE PILGRIM）

My heart is tuned to the quietness that the stillness of nature inspires.

大自然的宁静使我的内心平静。

——哈兹拉特·伊那亚特·可汗

（HAZRAT INAYAT KHAN）

Mon. Tue. Wed. Thur. Fri. Sat. Sun.

Date

年

月

日

Cheerfulness keeps up a kind of daylight in the mind, and fills it with a steady and perpetual serenity.

快乐使心灵保持光明，也让心灵充盈着平静安稳。

——约瑟夫·艾迪生

（JOSEPH ADDISON）

There are two days in the week about which and upon which I never worry. Two carefree days, kept sacredly free from fear and apprehension. One of the days is Yesterday.... And the other day I do not worry about is Tomorrow.

一周中有两天我从不烦恼，无忧无虑，自由自在：一天是昨天，另一天是明天。

——罗伯特·琼斯·伯德特

（OBERT JONES BURDETTE）

Mon. Tue. Wed. Thur. Fri. Sat. Sun.

Date

年

月

日

I am searching for that which every person seeks – peace and rest.

我在寻找所有人都追寻的——和平与安宁。

——但丁·阿利吉耶里

(DANTE ALIGHIERI)

Is it so small a thing to have enjoyed the sun, to have lived light in the spring, to have loved, to have thought, to have done?

享受过温暖的阳光，感受过和煦的春光，相爱过，思考过，奋斗过，这些难道是不值一提的小事？

——马修·阿诺德

（MATTHEW ARNOLD ）

Mon. Tue. Wed. Thur. Fri. Sat. Sun.

Date

年

月

日

Enjoy the little things, for one day you may look back and discover they were the big things.

珍惜微小的事物，因为有一天回首往事，你会发现，它们何其珍贵。

——选自《散播希望》

（FROM “SHARE THE HOPE”）

Mon. Tue. Wed. Thur. Fri. Sat. Sun.

Date

年

月

日

Difficulties along the way are opportunities in disguise; they reflect your expectations. Facing them with surrender helps you follow a more peaceful and perceptive life.

沿途的困难其实是机遇，它们反映了你的期望。承认失败能让你更平静从容地生活。

——卡尔·阿博特

（CARL ABBOTT）

Mon. Tue. Wed. Thur. Fri. Sat. Sun.

Date 年 月 日

Inside myself is a place where I live all alone, and that's where you renew your springs that never dry up.

在我心中有一处独处空间，那是我永不干涸的精神源泉。

——赛珍珠

（PEARL S. BUCK）

A small house will hold as much happiness as a big one.

小屋子容纳的幸福与大房子一样多。

——佚名

（AUTHOR UNKNOWN）

Mon. Tue. Wed. Thur. Fri. Sat. Sun.

Date

年

月

日

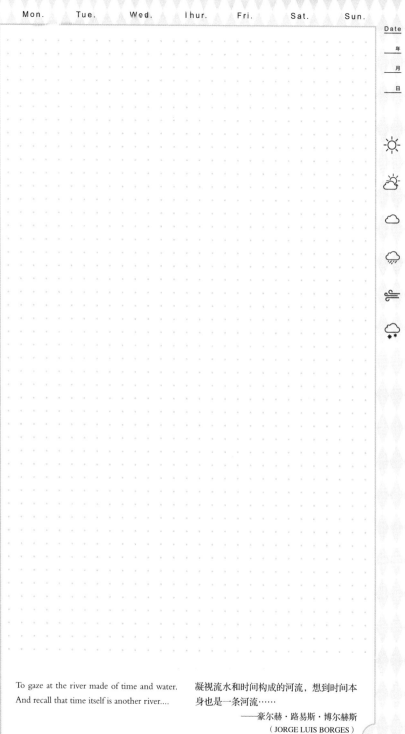

To gaze at the river made of time and water.
And recall that time itself is another river....

凝视流水和时间构成的河流，想到时间本身也是一条河流……

——豪尔赫·路易斯·博尔赫斯
（JORGE LUIS BORGES）

Sitting silently,
Doing nothing,
Spring comes,
And the grass grows by itself.

静坐，无事。春来，草长。

——松尾芭蕉
（BASHO）

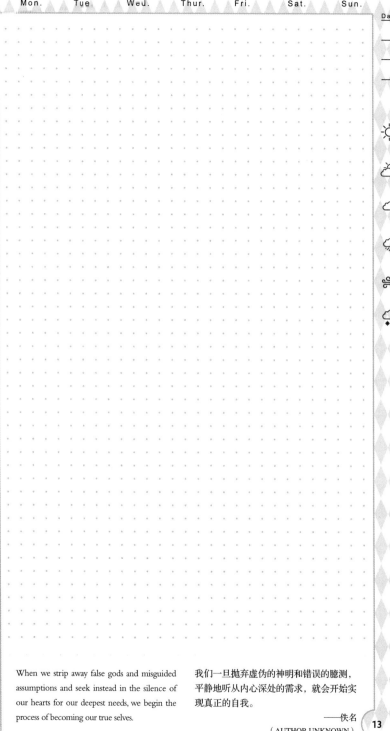

When we strip away false gods and misguided assumptions and seek instead in the silence of our hearts for our deepest needs, we begin the process of becoming our true selves.

我们一旦抛弃虚伪的神明和错误的臆测，平静地听从内心深处的需求，就会开始实现真正的自我。

——佚名

（AUTHOR UNKNOWN）

I am never less alone than when alone.

独处之时，我从不觉得太过孤独。

——大西庇阿

(SCIPIO AFRICANUS)

Mon. Tue. Wed. Thur. Fri. Sat. Sun.

Date

年

月

日

What makes a river so restful to people is that it doesn't have any doubt – it is sure to get where it is going, and it doesn't want to go anywhere else.

对于人们来讲，河流的宁静在于不惑，它知道自己的方向，也从未想过流向他方。

——哈尔·博伊尔

（HAL BOYLE）

The passing minute is every person's equal possession.

流逝的时间是人人平等的财富。

——马可·奥勒留

（MARCUS AURELIUS ）

Only in quietness can the infinity of wonder find you.

只有安静的时候，无限奇迹才会找上你。

——帕姆·布朗

（PAM BROWN）

The best day is... today!

最好的日子就是……今天！

——佚名

（AUTHOR UNKNOWN）

Mon. Tue. Wed. Thu. Fri. Sat. Sun.

Date

年

月

日

Happiness consists not in having much, but in being content with little.

幸福不在于拥有太多，而在于小小的满足。

——玛格丽特·嘉丁纳

（MARGUERITE GARDINER）

Wisdom can be possessed by the active but seems to favour those who lead the most simple lives.

充满活力的人也可能拥有智慧，但生性简朴的人往往更受智慧青睐。

——帕梅拉·达格代尔（PAMELA DUGDALE）

Decisions made in the heat of emotion may be brave, but not wise. Wait until the clamour has died away and in the calm you will see what must be done.

情绪激动的时候做出的决定也许大胆，却不明智。待喧嚣散去，回归平静，你会发现什么才是最明智的选择。

——帕姆·布朗（PAM BROWN）

Joy exists only in self-acceptance. Seek perfect acceptance, not a perfect life.

喜悦在于自我接纳。尝试完全接受自己，而不是追求完美的人生。

——佚名

（AUTHOR UNKNOWN）

How can the world know peace when we have lost the gift of stillness?

如果我们不能安静下来，世界又怎能变得宁静？

——玛雅 · V. 帕特尔
（MAYA V. PATEL）

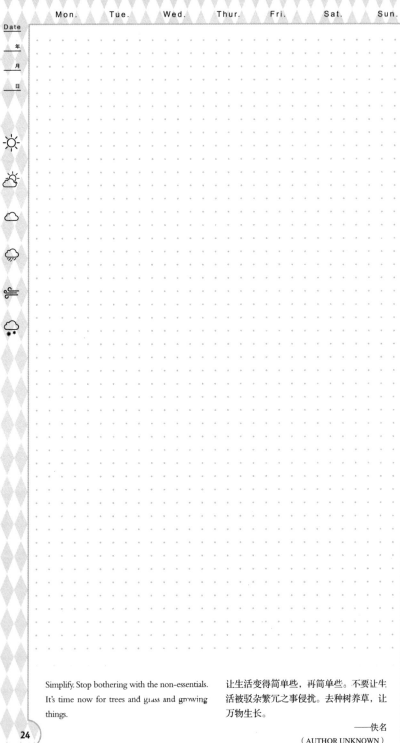

Simplify. Stop bothering with the non-essentials. It's time now for trees and grass and growing things.

让生活变得简单些，再简单些。不要让生活被驳杂繁冗之事侵扰。去种树养草，让万物生长。

——佚名

（AUTHOR UNKNOWN）

Date

年

月

日

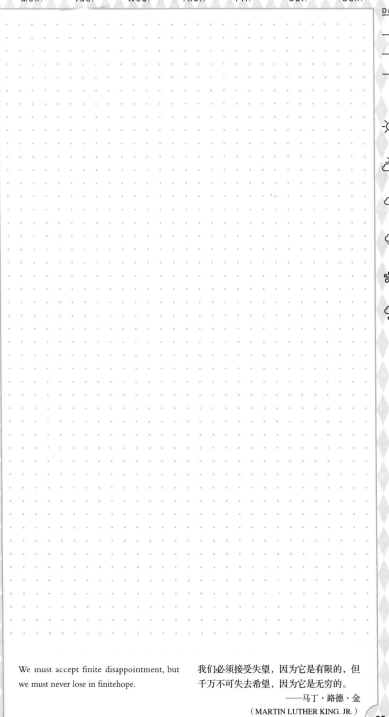

We must accept finite disappointment, but we must never lose in finitehope.

我们必须接受失望，因为它是有限的，但千万不可失去希望，因为它是无穷的。

——马丁·路德·金

（MARTIN LUTHER KING. JR.）

Drive out one by one the small obsessions that torment you. Silence desire, ambition and anxiety. Be still. Let the quietness fill you. Be one with every living thing.

放弃折磨你的细小执念。让欲望、野心和焦虑都安顿下来。静下来，让宁静填满你的身心，与万物合一。

——帕姆·布朗（PAM BROWN）

Just for today I will try to live through this day only, and not tackle my life's problems at once.

活在当下，行在今日。我将尽力活好今天，而不是立即去解决生活中的烦事。

——佚名（AUTHOR UNKNOWN）

Let go of any feeling of being tied to time. Flow with it and you will be in the right place at the right time doing the right thing.

抛开时间的枷锁。跟随它一道流逝，你自然会在正确的时间，到正确的地点，做正确的事情。

——艾琳·卡迪

（EILEEN CADDY）

Mon. Tue. Wed. Thur. Fri. Sat. Sun.

Date
年
月
日

The trail is beautiful. Be still.

这是一条幽美的小径。停住，静静欣赏。

——佚名

（ANONYMOUS）

Do not seek to have everything that happens happen as you wish, but wish for everything to happen as it actually does happen, and your life will be serene.

不要期待万事如你所愿，只要万事顺其自然，生活自然将宁静安详。

——爱比克泰德

（EPICTETUS）

Mon. Tue. Wed. Thur. Fri. Sat. Sun.

Date

年

月

日

If you want to be happy, be.

要是想快乐，就快乐吧。

——列夫·托尔斯泰

（LEO TOLSTOY）

The person who, casting off all desires, lives free from attachment; who is free from egoism and from the feeling that this or that is mine, obtains tranquillity.

摆脱所有欲望的人，不受外来事物牵绊；放弃利己主义和一切皆为我所有想法的人，心境会更加稳定。

——《博伽梵歌》

（THE BHAGAVAD GITA）

Mon. Tue. Wed. Thur. Fri. Sat. Sun.

Date 年 月 日

Perhaps this very instant is your time...your own, your peculiar, your promised and presaged moment, out of all moments forever.

也许这个时刻就是你的时刻……它是你自己的、独特的、应得的、注定的时刻，所有时刻独属于你。

——路易斯·博根
（LOUISE BOGAN）

Not wanting to be anyone or anything but what we are allows us to be where we are. No longer straining our sight, not dreaming of another world, we are freed to discover this one.

做自已，而不是想着变成什么其他人物，我们就能够随遇而安。不要再吃力地眺望远方，不要再梦想新世界，便能自由探索眼前的这个世界。

——迈克尔·亚当（MICHAEL ADAM）

Mon. Tue. Wed. Thur. Fri. Sat. Sun.

Date

年

月

日

An old man said: "Human care and worry and anxiety about the things of the body destroy the faculties of knowledge and expression in us, and leave us like unto a piece of dry wood."

一位老者说过："人类对肉体的关注、担心和焦虑摧毁了我们的知识和表达能力，让我们成了一块朽木。"

——帕拉第乌斯
（PALLADIUS）

The silence which vibrates around contains more sense of presence than sound.

环绕在周围的寂静远比声音蕴含着更多存在的意义。

——凯瑟琳·休伊特

（CATHERINE HEWITT）

I wish you joy and peace and deep contentment. And always, always, love.

愿你拥有幸福、安宁、满足。还有，永恒的爱。

——帕梅拉·达格代尔

（PAMELA DUGDALE）

You ask why I make my home in the mountain forest, and I smile, and am silent, and even my soul remains quiet: it lives in the other world which no one owns. The peach trees blossom. The water flows.

问余何意栖碧山，笑而不答心自闲。
桃花流水窅然去，别有天地非人间。

——李白

（LI PO）

Mon. Tue. Wed. Thur. Fri. Sat. Sun.

Date

年

月

日

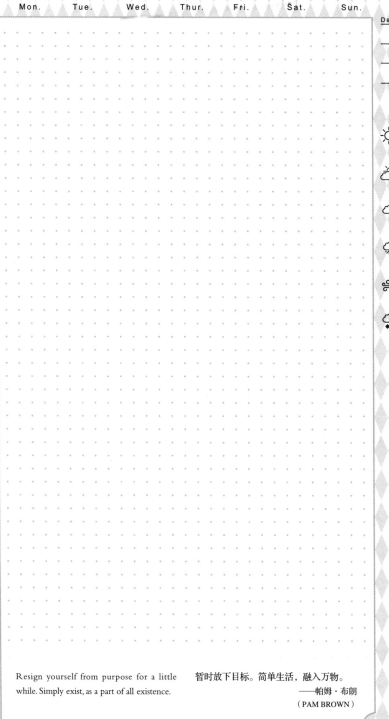

Resign yourself from purpose for a little while. Simply exist, as a part of all existence.

暂时放下目标。简单生活，融入万物。

——帕姆·布朗

（PAM BROWN）

My life... I began to realize, lacks this quality of significance and therefore of beauty, because there is so little empty space. The space is scribbled on; the time has been filled.

我逐渐意识到我的生活没有意义，也缺少美好，因为它没有些许空间。时间和空间都被胡乱地塞满了。

——安妮 · 默洛 · 林德伯格

（ANNE MORROW LINDBERGH）

It seems to me I'd like to go
Where bells don't ring, nor whistles blow,
Nor clocks don't strike, nor gongs sound,
And I'd have stillness all around.

我好想去一个地方，
没有铃声，没有哨声，
没有钟声，没有锣声，
我为寂静所环绕。

——尼克松·沃特曼

（NIXON WATERMAN）

Only when the mind has learned to be still can joy and beauty find a place.

只有学会静下心来，幸福和美好才能找到归宿。

——帕姆·布朗

（PAM BROWN）

My creed is this: Happiness is the only good. The place to be happy is here. The time to be happy is now.

我相信：幸福才是唯一的美好。幸福就在此地，幸福就在此时。

——罗伯特·英格索尔

（ROBERT INGERSOLL）

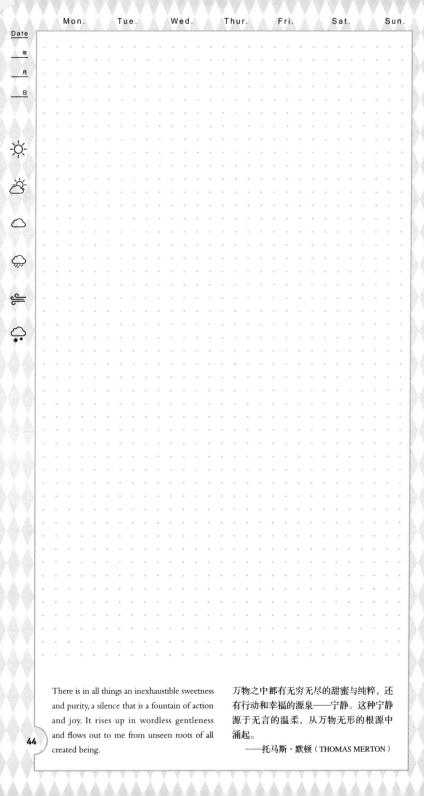

There is in all things an inexhaustible sweetness and purity, a silence that is a fountain of action and joy. It rises up in wordless gentleness and flows out to me from unseen roots of all created being.

万物之中都有无穷无尽的甜蜜与纯粹，还有行动和幸福的源泉——宁静。这种宁静源于无言的温柔，从万物无形的根源中涌起。

——托马斯·默顿（THOMAS MERTON）

The birds have vanished down the sky.
Now the last cloud drains away.
We sit together, the mountain and I,
until only the mountain remains.

众鸟高飞尽，孤云独去闲。
相看两不厌，只有敬亭山。

——李白
（LI PO）

We are a part of all that is. The branching trees find echoes in our veins. The spring that calls the buds to break and the swallows to return – waken us to joy. May you, today and always, share the wonder of the world.

我们都是万物的一部分。繁茂的树木在我们的血管间荡起回声。春暖花开燕子归来，呼唤我们享受幸福。愿你与我永远共享世界的美好。

——帕姆·布朗（PAM BROWN）

Mon. Tue. Wed. Thur. Fri. Sat. Sun.

Date

年

月

日

The lure of the distant and the difficult is deceptive. The great opportunity is where you are.

远方的诱惑和对困难的征服欲终究是虚幻的。最好的机会就在你的眼前。

——约翰·巴勒斯

（JOHN BURROUGHS）

The hours when the mind is absorbed by beauty are the only hours we live.

心灵被美吸引之时，才是我们活着的时刻。
——理查德·杰弗里斯
（RICHARD JEFFERIES）

Mon. Tue. Wed. Thur. Fri. Sat. Sun.

Date

年

月

日

When my heart is at peace the world is at peace.

当我的心平静的时候，世界就平静了。

——谚语

（PROVERB）

To be content with what we possess is the greatest and most secure of riches.

知足是最大且最可靠的财富。

——马库斯·图利乌斯·西塞罗

（MARCUS TULLIUS CICERO）

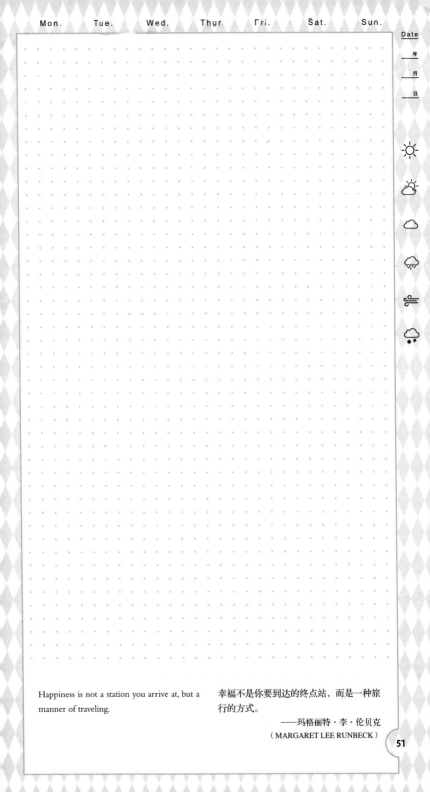

Happiness is not a station you arrive at, but a manner of traveling.

幸福不是你要到达的终点站，而是一种旅行的方式。

——玛格丽特·李·伦贝克（MARGARET LEE RUNBECK）

The best thinking has been done in solitude. The worst has been done in turmoil.

最好的灵感来自独处之时。最糟的主意产生于混乱之时。

——托马斯·爱迪生
（THOMAS EDISON）

There is only "now" and that holds the past, present and future.

只有“当下”才能把握住过去、现在和将来。

——乔伊斯·顾莲费尔

（JOYCE GRENFELL）

Mon. Tue. Wed. Thur. Fri. Sat. Sun.

Date

年

月

日

Each day the first day: Each day a life.

每一天都是余生的第一天，每一天都是新生活的开端。

——达格·哈马舍尔德

（DAG HAMMARSKJÖLD）

Perfect bliss grows only in the heart made tranquil.

内心的宁静，才能滋养心灵的幸福。

——印度教谚语

（HINDU PROVERB）

The miracle is not to fly in the air, or to walk on the water, but to walk on the earth.

奇迹不是在稀薄的空气中或水面上行走，而是在大地上行走。

——谚语
（PROVERB）

Mon. Tue. Wed. Thur. Fri. Sat. Sun.

Date

年

月

日

Silence is more musical than any song.

大音希声，无声是最美妙的乐曲。

——克里斯蒂娜・罗塞蒂

（CHRISTINA ROSSETTI）

God grant me serenity to accept the things I cannot change, courage to change the things I can, and wisdom to know the difference.

上帝请赐给我平静的心，让我接受无法改变的事实；请赐我勇气，改变能改变的事情；赐给我智慧，来分辨两者之间的区别。

——威廉·詹姆斯

（WILLIAM JAMES）

Mon. Tue. Wed. Thur. Fri. Sat. Sun.

Date

年

月

日

Take the gentle path.

走徐缓的路，无须刻意求险。

——乔治·赫伯特

（GEORGE HERBERT）

The person is richest who is content with the least; for content is the wealth of nature.

知足者最为富有，因为知足本身就是财富。

——苏格拉底

（SOCRATES）

Mon. Tue. Wed Thur. Fri. Sat. Sun.

Date
年
月
日

To sit in the shade on a fine day and look upon verdure is the most perfect refreshment.

放松心情最完美的方式，莫过于晴好的日子里坐在树下，看着郁郁葱葱一片绿色。

——简·奥斯汀

（JANE AUSTEN）

What is Life? It is the flash of a firefly in the night. It is the breath of a buffalo in the winter time. It is the little shadow which runs across the grass and loses itself in the sunset.

生命是什么？它是黑夜里萤火虫的光芒，是冬日里水牛的呼吸声，是掠过草地最终消失在日落中的阴影。

——克劳福特
（CROWFOOT）

The secret of the future is here in the present. If you pay attention to the present, you can improve upon it. And, if you improve on the present, what comes later will also be better....Each day, in itself, brings with it an eternity.

未来的秘密就在当下。如果你留心当下，就会不断去改进它；要是当下更好，未来也会更加美好……每一天本身都是无穷无尽的。

——保罗·科埃略
（PAULO COELHO）

Life's sweetest things are the quietest things...
A happy life consists of tranquillity of mind.

生活中最甜蜜的便是最安静的事物……幸福的生活来源于心灵的平静。

——马库斯·图利乌斯·西塞罗（CICERO）

Mon. Tue. Wed. Thur. Fri. Sat. Sun.

Date

年

月

日

Take the time to come home to yourself every day.

每天回到家，花点时间回归自我。

——罗宾·葛萨姜

（ROBIN CASARJIAN）

Silence is deep as eternity; speech, shallow as time.

沉默深沉永恒，言语肤浅易逝。

——托马斯·卡莱尔

（THOMAS CARLYLE）

Mon. Tue. Wed. Thur. Fri. Sat. Sun.

Date
年
月
日

Be able to be alone. Lose not the advantage of solitude, and the society of thyself.

要有独处的能力。不要放弃独处的好处，也不要放弃拥有自己的世界。

——托马斯·布朗爵士

（SIR THOMAS BROWNE）

Do not let trifles disturb your tranquillity of mind... life is too precious to be sacrificed for the non-essential and transient....

不要让琐事扰乱你内心的安宁……生命苦短，不该浪费在转瞬即逝的小事上。

——葛弗尔·克莱什尔

（GRENVILLE KLEISER）

Mon. Tue. Wed. Thur. Fri. Sat. Sun.

Date 年 月 日

I wish you the happiness of letting the past go – and finding new beginnings.

愿你放下过去，重新开始，找到幸福。

——帕姆·布朗

（PAM BROWN）

A happy life is made up less of great events than little, lovely moments.

幸福生活与其说来自许多丰功伟绩，莫若说是由微小美妙的瞬间组成的。

——玛雅·V. 帕特尔

（MAYA V. PATEL）

Mon. Tue. Wed. Thur. Fri. Sat. Sun.

Date

年

月

日

Nothing can bring you peace but yourself.

除了自己，谁都给不了你安宁。

——拉尔夫·沃尔多·爱默生

（RALPH WALDO EMERSON）

Be still. Let the mind go free to rediscover lost and lovely things. Empty your hands and you will be filled with peace.

安静！让心灵重新发现失去的美好。放下手中的一切，就能收获满满的安宁。

——帕姆·布朗

（PAM BROWN）

Simplicity, clarity, singleness: these are the attributes that give our lives power and vividness and joy.

要变得简单、纯粹、专一。它们赐予生命力量、生机和欢乐。

——理查德·霍洛威

（RICHARD HALLOWAY）

Take the breath of the new dawn and make it part of you. It will give you strength.

黎明到来时，做个深呼吸，让这成为你生活的一部分。它将赋予你力量。

——霍皮人

（HOPI）

Mon. Tue. Wed. Thur. Fri. Sat. Sun.

Date

年

月

日

Serenity is neither frivolity, nor complacency, it is the highest knowledge and love, it is the affirmation of all reality being awake at the edge of all deeps and abysses. Serenity is the secret of beauty and the real substance of all art.

平静不是轻率，也不是自满，而是至高的学问和爱，是身临深渊仍对事实保持清醒的肯定。平静是美的秘密，是所有艺术的本质。

——赫尔曼·黑塞（HERMANN HESSE）

Mon. Tue. Wed. Thur. Fri. Sat. Sun.

Date

年

月

日

In the bustle of life.In the pressure of decisions, peace has become a luxury. Take it when it comes, and cherish it. It gives you the time to breathe. It gives you rest and hope and life.

在喧嚣的生活中和选择的压力下，宁静是一种奢侈。当它到来的时候，接受它，珍惜它。它给予你呼吸的时间，让你休憩，给你希望和生命。

——帕姆·布朗（PAM BROWN）

Date

年

月

日

The happiness of life is made up of minute fractions – the little, soon-forgotten charities of a kiss, a smile, a kind look, a heartfelt compliment.

幸福的生活由细碎的片段组成：很快被遗忘的轻轻的一个吻，脸庞上的笑容，和善的目光和真心的赞美。

——塞缪尔·泰勒·柯勒律治

（SAMUEL TAYLOR COLERIDGE）

Mon. Tue. Wed. Thur. Fri. Sat. Sun.

Date

年

月

日

We ascribe beauty to that which is simple; which has no superfluous parts; which exactly answers its ends.

我们把美归结于简单的事物。这种事物没有繁文缛节，因而能精确地道出它的目的。

——拉尔夫·沃尔多·爱默生

（RALPH WALDO EMERSON）

Mon. Tue. Wed Thur. Fri. Sat. Sun.

Date

年

月

日

We need perfect simplicity with regard to ourselves, perfect contentment with all that comes our way, perfect peace of mind in utter self-forgetfulness.

纯粹简单地面对自己，知足常乐地面对困难。在完全忘却自我中获得心灵的完美平静。

——古迪尔大主教

（ARCHBISHOP GOODIER）

It is now and in this world that we must live.

我们必须活在当下，生活在这个世界上。

——安德烈·纪德

（ANDRÉ GIDE）

Nothing is so strong as gentleness; nothing so gentle as real strength.

没有什么比温柔更强大，没有什么比真正的力量更温柔。

——圣弗朗西斯·德·萨勒

（STFRANCIS DE SALES）

Mon. Tue. Wed. Thur. Fri. Sat. Sun.

Date

年

月

日

People miss their share of happiness, not because they never found it, but because they didn't stop to enjoy it.

人们错过幸福，不是因为没有发现它，而是因为没有停下来享受它。

——威廉·费瑟

（WILLIAM FEATHER）

I wish you quiet sleep, dreams of meadows, deep in flowers and grass, of oceans calm and flecked with silver, of islands hushed by gentle waves....

愿你沉沉入睡，梦见草坪上遍地花草，梦见大海宁静，波光粼粼，岛屿为微波抚慰，悄然无声。

——帕姆 · 布朗
（PAM BROWN）

Each moment of the year has its own beauty, a picture which was never seen before, and which shall never be seen again.

一年中的每个时刻都有独特的美，这种美从没出现过，也永远不会重现。

——拉尔夫·沃尔多·爱默生

（RALPH WALDO EMERSON）

Never be afraid to sit awhile and think.

永远不要害怕静坐思考。

——洛琳·汉斯伯里

（LORRAINE HANSBERRY）

To be sensual is to respect and rejoice in the force of life itself, and to be present in all that one does, from the effort of loving to the breaking of bread.

感受生命，就要尊重生命的力量，为它感到欢喜，参与生命中的一切，大到努力去爱，小到吃一餐饭。

——菲尔·加兰

（PHYL GARLAND）

Calm is a clear well that you may draw from whenever you have need.

平静是一口井，无论何时需要，都取之不竭。

——玛雅 · V. 帕特尔

（MAYA V. PATEL）

Silence is necessary. I have never ceased to mount guard around my silence. I have defended it against my enemies – (that is nothing); I have defended it especially against my friends.

沉默是必要的。我从未停止过对我的沉默严加防卫。我防卫它不受敌人的侵害——（这不算什么）；我尤其要防卫它不受朋友的侵害。

——罗曼·罗兰（ROMAIN ROLLAND）

At the centre of the most turbulent heart there is a place of peace, a place beyond time that cannot be touched by change or loss. No tumult can disturb that quietness, no shadow can dim that light. Here in this stillness is rest and healing. Nothing we suffer, nothing that we fear, can damage its perfection.

激荡不安的内心深处有一个平静的角落，超越时间的束缚，不会改变，也不会被失落撼动。任何混乱都不能打破这种平静，没有阴影能遮挡它的光芒。这种平静供你休憩，治愈伤痛。我们的痛苦与恐惧都无法打破这种极致的平静。

——帕姆·布朗（PAM BROWN）

Learn to be silent. Let your mind listen and absorb.

学会安静。让你的头脑学会倾听并吸纳理解。

——毕达哥拉斯

（PYTHAGORA）

A person of understanding remains silent.

明理的人，沉默不语。

——《圣经》

（THE BIBLE）

Mon. Tue. Wed. Thur. Fri. Sat. Sun.

Date

年

月

日

Life is the first gift, love is the second, and understanding the third.

生命是第一重礼赞，爱是第二重，理解是第三重。

——玛吉·皮尔斯
（MARGE PIERCY）

Date

年

月

日

He is happiest, be he king or peasant, who finds peace in his home.

无论是国王还是农夫，只要家庭和睦，就是世上最幸福的人。

——约翰·沃尔夫冈·冯·歌德

（JOHANN WOLFGANG VON GOETHE ）

Mon. Tue. Wed. Thur. Fri. Sat. Sun.

Date

年

月

日

Every day is a good day.

日日是好日。

——云门文偃禅师

（YUN MEN）

Mon. Tue. Wed. Thur. Fri Sat. Sun.

Date

年

月

日

I am beginning to learn that it is the sweet, simple things of life which are the real ones after all.

我开始意识到，生活中简单甜蜜的事物才是最实在的。

——劳拉·英格尔斯·怀尔德
（LAURA INGALLS WILDER）

Mon. Tue. Wed. Thur. Fri. Sat. Sun.

Date

年

月

日

My favorite piece of music is the one we hear all the time if we are quiet.

我最喜爱的乐曲，就是我们安静的时候总会听到的那首。

——约翰·凯奇

（JOHN CAGE）

Mon. Tue. Wed. Thur. Fri. Sat. Sun.

Date

年

月

日

I love to be alone. I never found the companion that was so companionable as solitude.

我喜欢独自生活。独处是我最友好的伙伴。

——亨利·戴维·梭罗

（HENRY DAVID THOREAU）

Mon. Tue. Wed. Thur. Fri. Sat. Sun.

Date

年

月

日

Having spent the better part of my life trying to re-live the past or experience the future before it arrives, I have come to believe that in between these two extremes is peace.

我花了大半生的时间，去努力回到往昔或体验将来。现在我终于意识到，这两个极端之间是宁静。

——乔伊·弗勒德

（JOY FLOOD）

Mon. Tue. Wed. Thur. Fri Sat. Sun.

Date

年

月

日

Before me peaceful.
Behind me peaceful.
Under me peaceful.
Over me peaceful.
Around me peaceful.

眼前是宁静的，身后是宁静的，
脚下是宁静的，头顶也是宁静的。
在我四方，万物俱静。

——纳瓦霍经文
（NAVAJO PRAYER）

Mon. Tue. Wed. Thur. Fri. Sat. Sun.

Date

年

月

日

Today is always here....
Tomorrow never.

今天永远都在……
明天遥不可及。

——托妮 · 莫里森
（TONI MORRISON）

Mon. Tue. Wed. Thur. Fri. Sat. Sun.

Date

年

月

日

Look lovingly upon the present, for it holds the only things that are forever true. All healing lies within it because its continuity is real.

用爱的眼神去看当下这一刻吧！因为永恒真实之物只可能存在于这一刻。一切疗愈都蕴含其中，因为只有它才具有真实的延续性。

——《奇迹课程》（A COURSE IN MIRACLES）

Mon. Tue. Wed. Thur. Fri. Sat. Sun.

Date

年

月

日

Loud words fall resoundingly into nothingness. The right word even spoken quietly, will enlighten the world.

大声喊出的话语，只会成为回声，沦为虚无。正确的话语，即使温和地说出，也能点亮世界。

——《摩诃婆罗多》

（THE MAHABHARATA）

Mon. Tue. Wed. Thur. Fri. Sat. Sun.

Date

年

月

日

The greatest wealth is to live content with little, for there is never want where the mind is satisfied.

知足是最大的财富。知足的人，永不匮乏。

——卢克莱修

（LUCRETIUS）

Mon. Tue. Wed. Thur. Fri. Sat. Sun.

Date

年

月

日

Stay at home, my heart, and rest.

待在家里，让我的心灵得到安顿，身心彻底休憩。

——亨利·沃兹沃斯·朗费罗

（HENRY WADSWORTH LONGFELLOW）

Mon. Tue. Wed. Thur. Fri. Sat. Sun.

Date

年

月

日

The greatest wealth is contentment with a little.

容易知足是最大的财富。

——谚语

（PROVERB）

Mon. Tue. Wed. Thur. Fri. Sat. Sun.

Date

年

月

日

Leave behind the dark well of your little world, look upwards to the limitless sky....

把你微小世界的黑暗之井抛在脑后，抬头仰望浩瀚无垠的天空。

——拉宾德拉纳特·泰戈尔

（RABINDRANATH TAGORE）

We need time to dream, time to remember, and time to reach the infinite. Time to be.

我们需要时间去想象，需要时间去铭记，需要时间去追求无限和永恒。是时候了。

——格莱迪丝·泰伯

（GLADYS TABER）

Mon. Tue. Wed. Thur. Fri. Sat. Sun.

Date

年

月

日

The acknowledgement of impermanence holds within it the key to life itself.

承认生命无常，便掌握了开启人生的钥匙。

——史蒂芬·莱文

（STEPHEN LEVINE）

Date

年

月

日

One of the greatest sounds of them all – and to me it is a sound – is utter, complete silence.

对我来说，最美妙的声音之一就是极致的宁静。

——安德烈·哥斯特兰尼兹

（ANDRÉ KOSTELANETZ）

Mon. Tue. Wed. Thur. Fri. Sat. Sun.

Date

年

月

日

It is neither wealth nor splendour, but tranquility and occupation, which give happiness.

财富和名声都无法带来幸福，只有宁静和工作才可以。

——托马斯·杰斐逊

（THOMAS JEFFERSON）

Mon. Tue. Wed. Thur. Fri. Sat. Sun.

Date

年

月

日

Happiness comes of the capacity to feel deeply, to enjoy simply, to think freely....

幸福是一种能力：深切地感受，简单地享受，自由地思考……

——斯道姆·詹姆森

（STORM JAMESON）

Mon. Tue. Wed. Thur. Fri. Sat. Sun.

Date

年

月

日

From serenity comes gentleness, comes reassurance, comes lasting strength.

宁静带来温柔，带来安慰，带来持久的力量。

——帕姆·布朗

（PAM BROWN）

Mon. Tue. Wed. Thur Fri. Sat. Sun.

Date

年

月

日

If only I may grow: firmer, simpler – quieter, warmer.

希望我能更坚定、更简单、更平静、更温暖，这该有多好啊。

——达格·哈马舍尔德

（DAG HAMMARSKJÖLD）

Mon. Tue. Wed. Thur. Fri. Sat. Sun.

Date

年

月

日

One today is worth two tomorrows.

一个今日胜过两个明天。

——本杰明·富兰克林

（BENJAMIN FRANKLIN）

Mon. Tue. Wed. Thur. Fri. Sat. Sun.

Date

年

月

日

Never has the earth been so lovely nor the sun so bright, as today....

世界从未像今天这样可爱，阳光也从未如今天这般明媚……

——尼基纳皮酋长

（CHIEF NIKINAPI）

Mon. Tue. Wed. Thur. Fri. Sat. Sun.

Date

年

月

日

Happiness is the flower of a long inner life of joy and contentment; it tells of peaceful hours and days on the sunniest heights of our soul.

幸福是一朵花儿，从长期喜悦满足的内在生活中绽放，在灵魂充满阳光的山巅诉说着宁静的岁月。

——莫里斯·梅特林克

（MAURICE MAETERLINCK）

Mon. Tue. Wed. Thur. Fri. Sat. Sun.

Date

年

月

日

And well the tiny things of earth repay the watching eye.

是的，地球上的小生命也值得我们去注视。

——艾比内泽·艾略特

（EBENEZER ELLIOTT）

Mon. Tue. Wed. Thur. Fri. Sat. Sun.

Date

年

月

日

Happiness is when what you think, what you say, and what you do are in harmony.

当你的所想、所言、所做达成和谐之时，幸福就会适时出现。

——圣雄甘地

（MAHATMA GANDHI）

Mon. Tue. Wed. Thur. Fri. Sat. Sun.

Date

年

月

日

Change is an easy panacea. It takes character to stay in one place and be happy there.

改变是很容易的灵丹妙药。它只需要在某处停留并感到快乐，这需要人格的力量。

——伊丽莎白·克拉克·邓恩

（ELIZABETH CLARKE DUNN）

Mon. Tue. Wed. Thur. Fri. Sat. Sun.

Date

年

月

日

The words the happy say are paltry melody.
But those the silent feel are beautiful.

快乐之人所能言说的只是寻常的旋律，但沉默之人所感受到的却是天籁之音。

——艾米莉·狄金森
（EMILY DICKINSON）

Date
年
月
日

An inch of time is an inch of gold: Treasure it. Appreciate its fleeting nature.

一寸光阴一寸金。珍惜光阴，感受光阴转瞬即逝的本质。

——李清云
（LOY CHING-YUEN）

Mon. Tue. Wed. Thur. Fri. Sat. Sun.

Date

年

月

日

The quieter you become the more you can hear.

越安静，你能听到的就越多。

——拉姆·达斯

（BABA RAM DASS）

Mon. Tue. Wed. Thur. Fri. Sat. Sun

Date

年

月

日

Don't hurry, don't worry. You're only here for a short visit. So be sure to stop and smell the flowers.

别行色匆匆，别愁眉苦脸。人生苦短，务必稍作停歇，嗅一嗅花香。

——华特・赫根

（WALTER HAGEN）

Mon. Tue. Wed. Thur. Fri. Sat. Sun.

Date

年

月

日

I am alone with the beating of my heart.

沉思空幽寂，岁月已徂征。

——刘基

（LUI CHI）

Mon. Tue. Wed. Thur. Fri. Sat. Sun.

Date

年

月

日

If you can spend a perfectly useless afternoon in a perfectly useless manner, you have learned how to live.

要享受悠闲的生活只要一种艺术家的性情，在一种全然悠闲的情绪中，去消遣一个闲暇无事的下午。

——林语堂

（LIN YUTANG）

Mon. Tue. Wed. Thur. Fri. Sat. Sun.

Date

年

月

日

Nothing is worth more than this day.

没有什么比今天更珍贵。

——约翰·沃尔夫冈·冯·歌德

（JOHANN WOLFGANG VON GOETHE）

Date

年

月

日

May peace and peace and peace be everywhere. 愿和平、和平与和平布满人间。

——《奥义书》

（THE UPANISHADS）

As you simplify your life, the laws of the universe will be simpler; solitude will not be solitude, poverty will not be poverty, nor weakness weakness.

当你让生活变得简单，宇宙的法则也会变得简单；孤独不再是孤独，贫穷不再是贫穷，弱点也不再是弱点。

——亨利·戴维·梭罗（HENRY DAVID）

The greatest revelation is stillness.

不欲以静，天下将自定。

——老子

（LAO–TZU）

Mon. Tue. Wed. Thur. Fri. Sat. Sun.

Date

年

月

日

True beauty must come, must be grown, from within....

真正的美必须来自内心，从内心生长……

——拉尔夫·沃尔多·川恩

（RALPH W. TRINE）

Mon. Tue. Wed. Thur. Fri. Sat. Sun.

Date

年

月

日

It is tranquil people who accomplish much.

心境平静的人才能成就伟业。

——亨利·戴维·梭罗

（HENRY DAVID THOREAU）

Mon. Tue. Wed. Thur. Fri. Sat. Sun.

Date

年

月

日

To a mind that is still the whole universe surrenders.

汝游心于淡……而天下治矣。

——庄子

（CHUANG TZU）

Date

年

月

日

Do not say, "It is morning,"and dismiss it with a name of yesterday. See it for the first time as a newborn child that has no name.

请不要说，“这是清晨”，然后用一个昨天的名字将它打发。把它看成一个初次相遇的无名的新生儿吧。

——拉宾德拉纳特·泰戈尔

（RABINDRANATH TAGORE）

Mon. Tue. Wed. Thur. Fri. Sat. Sun.

Date

年

月

日

Much silence has a mighty noise.

有多静就有多响。

——斯瓦西里谚语

（SWAHILI）

Mon. Tue. Wed. Thur. Fri Sat. Sun.

Date

年

月

日

I love tranquil solitude, and such society as is quiet, wise, and good.

我爱恬静的幽居，也爱沉静、明智、善良的群体。

——珀西·比希·雪莱

（PERCY BYSSHE SHELLEY）

Mon. Tue. Wed. Thur. Fri. Sat. Sun.

Date

年

月

日

Here will we sit and let the sounds of music creep in our ears: soft stillness and the night become the touches of sweet harmony.

我们坐在这里，让音乐之声潜入耳内：温柔的寂静和夜色足以衬托音乐的甜美。

——威廉·莎士比亚

（WILLIAM SHAKESPEARE）

Mon. Tue. Wed. Thur. Fri. Sat. Sun.

Date

年

月

日

Peace is when time doesn't matter as it passes by.

平静就是时间的流逝不再重要。

——玛丽亚·谢尔

（MARIA SCHELL）

Mon. Tue. Wed. Thur. Fri. Sat. Sun.

Date

年

月

日

To be without some of the things you want is an indispensable part of happiness.

幸福有个必不可少的条件，就是缺少一些你想要的东西。

——伯特兰·罗素

（BERTRAND RUSSELL）

Date

年

月

日

Enjoy life, employ life. It flits away and will not stay.

尽情地享受生活，感受生活的美好。否则它将快速流逝，不会久留。

——谚语

（PROVERB）

Mon. Tue. Wed. Thur. Fri. Sat. Sun.

Date

年

月

日

While I would not have missed yesterday, I have no desire to go back and live it over. For me, there is only the great today and the promise of tomorrow.

虽然我不曾错过昨日，但我从未渴望过昨日重现。对我来说，只有美好的今日和明日的吉兆。

——璧克馥

（MARY PICKFORD）

Mon. Tue. Wed. Thur. Fri. Sat. Sun.

Date

年

月

日

To become a happy person, have eyes that see romance in the commonplace, a child's heart, and spiritual simplicity.

要做一个快乐的人，要有一双能在平凡之处看到浪漫的慧眼，还要有一颗童心和朴素的精神。

——诺曼·文森特·皮尔

（NORMAN VINCENT PEALE）

Mon. Tue. Wed. Thur. Fri. Sat. Sun.

Date

年

月

日

Happiness sneaks in through a door you didn't know you left open.

幸福会从你不经意打开的门中悄然而至。

——约翰·巴里莫尔

（JOHN BARRYMORE）

Mon. Tue. Wed. Thur. Fri. Sat. Sun.

Date

年

月

日

I suppose the moments one most enjoys are moments – alone – when one unexpectedly stretches something inside you that needs stretching.

我想一个人最享受的时刻就是独处的时光。那时，人能不经意地舒展内心需要释放的东西。

——乔治娅·奥·吉弗

（GEORGIA O'KEEFFE）

Mon. Tue. Wed. Thur. Fri. Sat. Sun.

Date

年

月

日

Peace is the fairest form of happiness.

平静是最美好的幸福。

——威廉·埃勒里·钱宁

（WILLIAM ELLERY CHANNING）

Date

年

月

日

The contented man, even though poor, is happy; the discontented man, even though rich, is sad.

知足者常乐，贪婪者常忧。

——智慧

（WISDOM）

Mon. Tue. Wed. Thur. Fri. Sat. Sun.

Date

年

月

日

Quieten your mind. Quieten your heart. Let peace enfold you.

让你的头脑静下来，你的内心静下来，让和平围绕在你周围。

——帕姆·布朗

（PAM BROWN）

Mon. Tue. Wed. Thur. Fri. Sat. Sun.

Date
年
月
日

Listen! Or your tongue will make you deaf.

要多倾听！多话将使你失聪。

——切罗基谚语

（CHEROKEE SAYING）

Mon. Tue. Wed. Thur. Fri. Sat. Sun.

Date

年

月

日

Nature is painting for us, day after day, pictures of infinite beauty if only we have eyes to see them....

只要我们有一双慧眼，就能看到大自然日复一日地为我们绘制着无限的美丽……

——约翰·罗斯金

（JOHN RUSKIN）

Date

年

月

日

Poor and content is rich, and rich enough.

人能安贫就是富有。

——威廉·莎士比亚

（WILLIAM SHAKESPEARE）

Mon. Tue. Wed. Thur. Fri. Sat. Sun.

Date

年

月

日

By keeping quiet, remaining attentive, and, hand in hand with that, by accepting reality – taking things as they are, and not as I wanted them to be – by doing all this, rare knowledge has come to me, and rare powers as well.

保持安静，保持专注，与此同时，接受现实——接受事物本来的样子，而不是我想要的样子——做到这些，我获得了珍贵的知识，也获得了罕见的力量。

——病人写给欧内斯特·荣格的信

（A PATIENT'S LETTER TO ERNEST JUNG）

Date

年

月

日

There is silence, there is peace, there is certainty. Quieten your mind.... You hold them within you.

这里有平静、安宁和安稳。静下心来……将它们珍藏于心中。

——帕姆·布朗

（PAM BROWN）

Mon. Tue. Wed. Thur. Fri. Sat. Sun.

Date

年

月

日

Yesterday is but a dream and tomorrow only a vision, but today well-lived makes every yesterday a dream of happiness and every tomorrow a vision of hope.

昨日只是一场幻梦，明日不过是一幅愿景；但只要过好今日，每个昨日都将化为美梦，每个明日都将充满希望的图景。

——梵语（SANSKRIT）

Date

年

月

日

Today, today, today. Bless us... and help us grow.

今日，今日，今日。请祝福我们……帮助我们成长。

——犹太新年礼拜仪式

（FROM THE ROSH HASHANAH LITURGY）

Mon. Tue. Wed. Thur. Fri. Sat. Sun.

Date

年

月

日

Every moment there is news coming out of silence.

每时每刻都有新闻打破沉寂。

——赖内·马利亚·里尔克

（RAINER MARIA RILKE）

Date

年

月

日

All our miseries derive from not being able to sit quiet in a room alone.

我们所有的苦恼，都源于无法独自一人安静地坐在房间里。

——布莱士·帕斯卡

（BLAISE PASCAL）

Mon. Tue. Wed. Thur. Fri. Sat. Sun.

Date

年

月

日

The best way to secure future happiness is to be as happy as is rightfully possible today.

确保未来快乐的最好方法，就是尽可能愉悦地度过今天。

——查尔斯· W. 艾略特

（CHARLES W. ELIOT）

Mon. Tue. Wed. Thur. Fri. Sat. Sun.

Date

年

月

日

Peace is inevitable to those who offer peace.

对于那些分享平静的人来说，平静是天经地义的。

——《奇迹课程》

（A COURSE IN MIRACLES）

Mon. Tue. Wed. Thur. Fri. Sat. Sun.

Date

年

月

日

We are always getting ready to live, but never living.

我们总在处于活着的状态，却从来没真正生活过。

——拉尔夫·沃尔多·爱默生

（RALPH WALDO EMERSON）

Mon. Tue. Wed. Thur. Fri. Sat. Sun.

Date

年

月

日

Silence is not a thing we make; it is something into which we enter. It is always there.

寂静不是我们制造的一种效果，而是我们进入的一种状态。它一直在那里。

——玛丽贝尔嬷嬷

（MOTHER MARIBEL）

Mon. Tue. Wed. Thur. Fri. Sat. Sun.

Date

年

月

日

Deep peace of the running wave to you.
Deep peace of the flowing air to you.
Deep peace of the quiet earth to you.

愿滚滚的波浪为您带来深沉的安宁。
愿流动的空气为您带来深沉的平静。
愿沉默的大地为您带来深沉的宁静。

——菲奥纳·麦克劳德
（FIONA MACLEOD）

Date

年

月

日

Life is a succession of moments. To live each one is to succeed.

生活是由一连串瞬间组成的，过好每个瞬间就是成功。

——科里塔·肯特

（CORITA KENT）

Mon. Tue. Wed. Thur. Fri. Sat. Sun.

Date

年

月

日

Normal day, let me be aware of the treasure you are. Let me learn from you, love you, bless you before you depart. Let me not pass you by in quest of some rare and perfect tomorrow. Let me hold you while I may, for it may not always be so.

平常的日子，让我意识到你是什么样的珍宝。让我向你学习，爱你，在你离开前祝福你。让我不要为了追寻罕见而完美的明天错过你。让我在可以拥抱你的时候拥抱你，因为这或许无法长久。

——玛丽·琼·伊里翁（MARY JEAN IRION）

Yes, to become simple and live simply, not only within yourself but also in your everyday dealings. Don't make ripples all around you, don't try so hard to be interesting, keep your distance, be honest, fight the desire to be thought fascinating by the outside world.

是的，变得简单，就过得简单，不只发自内心，而且在日常生活中如此。不要在周围激起涟漪，不要竭力表现得有趣，与人保持距离，坦诚面对世界，抵抗吸引外界目光的欲望。

——艾提·海勒申（ETTY HILLESUM）

Mon. Tue. Wed. Thur. Fri. Sat. Sun.

Date

年

月

日

I expand and live in the warm day like corn and melons.

我在温暖的日子里茁壮成长，就像玉米和西瓜一样。

——拉尔夫·沃尔多·爱默生

（RALPH WALDO EMERSON）

Mon Tue. Wed. Thur. Fri. Sat. Sun.

Date

年

月

日

Let your soul stand cool and composed before a million universes.

让你的灵魂镇定沉着地面对天地万物。

——沃尔特·惠特曼

（WALT WHITMAN）

Mon. Tue. Wed. Thur. Fri. Sat. Sun.

Date

年

月

日

The service of the fruit is precious, the service of the flower is sweet, but let my service be the service of the leaves in its shade of humble devotion.

果实的事业是珍贵的，花朵的事业是甜美的，但是让我做叶的事业吧，叶是谦逊的，专心地垂着绿荫的。

——罗宾德拉纳特・泰戈尔

（RABINDRANATH TAGORE）

Mon Tue. Wed. Thur. Fri. Sat. Sun.

Date

年

月

日

Life is so short we must move very slowly.

生命苦短，我们必须得慢慢来。

——泰国谚语

（THAI PROVERB）

Mon. Tue. Wed. Thur. Fri. Sat. Sun.

Date

年

月

日

There is a quietness, a contentment, at the heart of all that is. If only you allow yourself to find it.

一切的中心都有一种宁静，一种满足。你只要允许自己找到它。

——帕姆 · 布朗

（PAM BROWN）

Mon. Tue. Wed. Thur. Fri. Sat. Sun.

Date

年

月

日

I do not fear tomorrow, for I have seen yesterday – and I love today.

我不畏惧明天，因为我已经见过了昨天——并且我热爱今天。

——威廉·艾伦·怀特

（WILLIAM ALLEN WHITE）

Happiness is not about being immortal nor having food or rights in one's hand. It's about having each tiny wish come true, or having something to eat when you are hungry or having someone's love when you need love.

幸福，不是长生不老，不是大鱼大肉，不是权倾朝野。幸福是每一个微小的生活愿望达成。当你想吃的时候有得吃，想被爱的时候有人来爱你。

——《飞屋环游记》(Up)

Mon. Tue. Wed. Thur. Fri. Sat. Sun.

Date

年

月

日

The one who knows that enough is enough will always have enough.

故知足之足，常足矣。

——老子

（LAO -TZU）

Mon. Tue. Wed. Thur. Fri. Sat. Sun.

Date

年

月

日

Peace starts with a smile.

平静从微笑开始。

——特蕾莎修女

（MOTHER TERESA）

Mon. Tue. Wed. Thur. Fri. Sat. Sun.

Date

年

月

日

Fear not for the future; weep not for the past. 不要为未来担忧，不要为过去悲泣。

——珀西·比希·雪莱

（PERCY BYSSHE SHELLEY）

Mon. Tue. Wed. Thur. Fri. Sat. Sun.

Date

年

月

日

Yesterday is history, tomorrow is a mystery, but today is a gift, that is why it's called the present.

昨日已成历史，明日尚未可知，今天是上天赐予我们的礼物，这就是为什么我们称之为现在。

——《功夫熊猫》

（KUNG FU PANDA）

True joy is serene.

平静是真正的快乐。

——塞内加

（SENECA）

Mon. Tue. Wed. Thur. Fri. Sat. Sun.

Date

年

月

日

Fear less, hope more; eat less, chew more; whine less, breathe more; talk less, say more; love more, and all good things will be yours.

与其恐惧，不如怀抱希望；与其贪食，不如细细品味；与其抱怨，不如静心感受；与其空谈，不如言之有物。常怀爱人之心，必将好事不断。

——瑞典谚语

（SWEDISH PROVERB）

Mon. Tue. Wed. Thur. Fri. Sat. Sun.

Date

年

月

日

Whatever peace I know rests in the natural world, in feeling myself a part of it, even in a small way.

我所知晓的平静在与自然之间，在与将自己融为其中之一，即便只是微小的一部分。

——梅·萨藤

（MAY SARTON）

Mon. Tue. Wed. Thur. Fri. Sat. Sun.

Date

年

月

日

Yes, there is a Nirvana, it is in leading your sheep to a green pasture, and in putting your child to sleep, and in writing the last line of your poem.

是的，有一种涅槃。它在于将你的羊群领去绿色的牧场，哄你的孩子入睡，写下你诗句的最后一行。

——卡里·纪伯伦

（KAHLIL GIBRAN）

Mon. Tue. Wed. Thur. Fri. Sat. Sun.

Date

年

月

日

Solitude is the profoundest fact of the human condition. We are the only beings who know we are alone.

孤独是人类处境中最深刻的事实。我们是唯一知道自己独自存在的生物。

——奥克塔维奥·帕斯
（OCTAVIO PAZ）

Mon. Tue. Wed. Thur. Fri. Sat. Sun.

Date

年

月

日

Away from books, away from art, the day erased, the lesson done. Thee fully forth emerging, silent, gazing, pondering the themes thou lovest best.

远离书本，远离艺术，一天过去，课程结束。你才完全显露，寂静凝视，默想你最喜爱的主题。

——沃尔特·惠特曼

（WALT WHITMAN）

Mon. Tue. Wed. Thur. Fri. Sat. Sun.

Date

年

月

日

Lovely, lasting peace of mind! Sweet delight of human kind!

愉悦让人心灵的宁静更持久，甜蜜让人更快乐。

——托马斯·帕内尔

（THOMAS PARNELL）

Mon. Tue. Wed. Thur. Fri. Sat. Sun.

Date

年

月

日

To have peace in one's soul is the greatest happiness.

心灵的宁静是最大的幸福。

——智慧

（WISDOM）

Date
年
月
日

The dream was always running ahead of one. To catch up, to live for a moment in unison with it, that was the miracle.

梦想永远跑在你前头。迎头追上，与它并驾齐驱，哪怕只有一瞬，便是奇迹。

——阿娜伊斯 · 宁
（ANAÏS NIN）

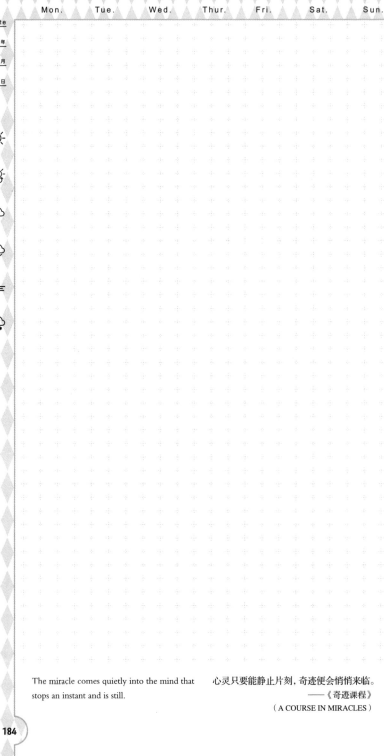

The miracle comes quietly into the mind that stops an instant and is still.

心灵只要能静止片刻，奇迹便会悄悄来临。

——《奇迹课程》

（A COURSE IN MIRACLES）

If there is to be any peace it will come through being, not having.

若宁静真的存在，那也源于顺其自然，而非靠拥有获得。

——亨利·米勒

（HENRY MILLER）

One should lie empty, open, choiceless as a beach – waiting for a gift from the sea.

人应该像海滩一样放空自己，不用选择地躺在平地上——等待来自海洋的馈赠。

——安妮·默洛·林德伯格

（ANNE MORROW LINDBERGH）

Mon. Tue. Wed. Thur. Fri. Sat. Sun.

Date

年

月

日

Life isn't a matter of milestones but of moments.

生活不在于某种方式的里程碑，而在于拥有每一个时刻。

——罗丝·菲茨杰拉德·肯尼迪

（ROSE FITZGERALD KENNEDY）

Solitude gives birth to the original in us, to beauty unfamiliar and perilous..

孤独孕育了我们的独创性，它是陌生且危险的美丽事物。

——托马斯·曼

（THOMAS MANN）

And silence like a poultice, comes to heal the blows of sound.

寂静如同一副膏药，可以治愈嘈杂声的打击。

——奥利弗·温德尔·霍姆斯

（OLIVER WENDELL HOLMES）

Within us should always be that peace which is forever listening and open-minded.

我们心中应当一直保持平和，它使我们能够认真倾听、心胸开阔。

——罗曼诺·古瓦尔迪尼
（ROMANO GUARDINI）

Mon. Tue Wed. Thur. Fri. Sat. Sun.

Date

年

月

日

All present beauty is the only beauty.

唯一的美就是现在的美。

——海伦·埃克斯利

（HELEN EXLEY）

Though we travel the world over to find the beautiful, we must carry it with us or we find it not.

就算我们踏遍全球寻找美好的事情，我们也必须将它随身带着，否则就会找不到。

——拉尔夫·沃尔多·爱默生

（RALPH WALDO EMERSON）

All else begone and leave me here alone – to tread this mist where earth and sky are one.

其余一切都请离去，独留我一人在天地合一处踩踏雾霭。

——安德鲁·杨（ANDREW YOUNG）

If the path be beautiful, let us not ask where it leads.

只要道路是美丽的，就不要问它通向何方。

——阿纳托尔·法朗士

（ANATOLE FRANCE）

Paradise is where I am.

天堂就是我身处之地。

——伏尔泰

（VOLTAIRE）

Mon. Tue. Wed. Thur. Fri. Sat. Sun.

Date

年

月

日

Once you plant deep the longing for peacc, confusion leaves of itself.

一种平怀，泯然自尽。

——僧璨

（SENG–T' SAN）

Mon. Tue. Wed. Thur. Fri. Sat. Sun.

Date

年

月

日

I'm filled with joy when the day dawns quietly over the roof of the sky.

当黎明悄然降临于天际，我的心中充满喜悦。

——《爱斯基摩人之歌》

（FROM “AN ESKIMO SONG”）

Mon. Tue. Wed. Thur. Fri. Sat. Sun.

Date

年

月

日

Our life is frittered away by detail.... Simplify, simplify.

我们的生活被琐事浪费了……简约，再简约生活。

——亨利·戴维·梭罗

（HENRY DAVID THOREAU）

Mon. Tue. Wed. Thur. Fri. Sat. Sun.

Date

年

月

日

There is no joy but calm.

无需欢愉，平静心智。

——阿尔弗雷德·丁尼生

（ALFRED, LORD TENNYSON）

Let my doing nothing when I have nothing to do become untroubled in its depth of peace, like the evening in the seashore when the water is silent.

当我没有什么事做时，便让我不做什么事。不受打扰地沉入安静的深处吧，一如海水沉默时海边的暮色。

——罗宾德拉纳特·泰戈尔

（RABINDRANATH TAGORE）

Mon. Tue. Wed. Thur. Fri. Sat. Sun.

Date

年

月

日

Yesterday has gone. Tomorrow may never come. There is only the miracle of this moment. Savor it. It is a gift.

昨日已逝，明日未到，唯一的奇迹在于此刻。请尽情品味此刻的一切，这是一份礼物。

——玛丽·斯蒂尔金德

（MARIE STILKIND）

Mon. Tue. Wed. Thur. Fri. Sat. Sun.

Date

年

月

日

In peace nothing so becomes a man as modest stillness and humility.

在和平时期，就要做个大丈夫，斯文平和还有谦逊都不可少。

——威廉·莎士比亚

（WILLIAM SHAKESPEARE）

Mon. Tue. Wed. Thur. Fri. Sat. Sun.

Date 年 月 日

When I am alone the flowers are really seen; I can pay attention to them. They are felt presences.

当我独自一人的时候，才能真正看见花朵，我得以注意它们，感受到它们的存在。

——梅·萨藤

（MAY SARTON）

Mon. Tue. Wed. Thur. Fri. Sat. Sun.

Date

年

月

日

We must live with beauty, without any straining effort to admire, quietly attentive, absorbent, until by degrees the beauty becomes one with us and alters our blood.

我们必须与美同在，不过分费力去赞美，而静静地注视和吸收。直到美与我们合一，和我们的血液融为一体。

——马克·卢瑟福（MARK RUTHERFORD）

Date

年

月

日

Each of us has... all the time there is. Those years, weeks, hours, are the sands in the glass running swiftly away.

我们每人都有……属于自己充足的时间。那些年、星期和小时，犹如沙子般飞速流逝。

——埃莉诺·罗斯福

（ELEANOR ROOSEVELT）

Our greatest experiences are our quietest moments.

我们最安静的时刻，就是我们最美好的经历。

——弗里德里希 · 威廉 · 尼采

（FRIEDRICH WILHELM NIETZSCHE）

Mon. Tue. Wed. Thur. Fri. Sat. Sun.

Date
年
月
日

Listen in deep silence. Be very still and open your mind.... Sink deep into the peace that waits for you beyond the frantic, riotous thoughts and sights and sounds of this insane world.

深深地静下来聆听吧！在极度的宁静中，开启你的心扉。……越过这疯狂世界的狂乱躁动的念头、景象和声音，向下沉潜到那静静等候你的平和中。

——《奇迹课程》

（A COURSE IN MIRACLES）

The morning sun, the new sweet earth and the great silence.

清晨的太阳，新鲜甜蜜的大地，还有极致的寂静。

——T.C. 麦克卢汉
（T.C. MCLUHAN）

Your life will never be fulfilled until you are happy here and now.

你的生命永远不会感到满足，直到你能在此时此地获得快乐。

——小肯·凯斯

（KEN KEYES, JR.）

You can destroy your now by worrying about tomorrow.

担心明天会毁了今天。

——詹妮丝·乔普林

（JANIS JOPLIN）

Mon. Tue. Wed. Thur. Fri. Sat. Sun.

Date

年

月

日

That it will never come again is what makes life so sweet.

生活之所以如此甜蜜，是因为它一去不复返。

——艾米莉·狄金森

（EMILY DICKINSON）

Mon. Tue. Wed. Thur. Fri. Sat. Sun.

Date

年

月

日

Life is a journey, not a destination; and happiness is not “there” but here; not tomorrow, but today.

生命就像一场旅行，不是以某个目的地为终点。快乐不在“那边”，而在这里；不在明天，而在今天。

——西德尼·格林伯格（SIDNEY GREENBERG）

There's the real danger of overlooking a very important day... today. For this is the place and the time for living. Let us live each day abundantly and beautifully while it is here.

真正的危险在于忽略人生中很重要的一天——今天。因为这是真正生活的时间和空间。让我们在今天没有逝去之前尽可能充实美好地度过吧。

——埃丝特·鲍德温·约克（ESTHER BALDWIN YORK）

Mon. Tue. Wed. Thur. Fri. Sat. Sun.

Date

年

月

日

Manifest plainness, embrace simplicity, reduce selfishness, have few desires.

见素抱朴，少私寡欲。

——老子

（LAO–TZU）

Commonly we stride through the out-of-doors too swiftly to see more than the most obvious and prominent things. For observing nature, the best pace is a snail's pace.

通常我们户外走路的步伐过于迅速，就看不见最明显最清楚的东西。要观察自然，最好就像蜗牛般走路。

——艾温 · 威 · 蒂尔
（EDWIN WAY TEAL）

Stay quiet; refuse nothing; flowers grow only because they tranquilly allow the sun's rays to reach them. You must do the same.

保持安静，什么也别拒绝。花儿生长只因为它们静静地让阳光照射自己。你也要如此。

——朱利安·库德纳男爵夫人

（BARONESS JULIANA KRUDENER）

The best person is like water. Water is good; it benefits all things and does not compete with them. It dwells in lowly places that all disdain.

上善若水，水善利万物而不争。处众人所恶，故几于道。

——老子

（LAO-TZU）

The creative soil of silence, where can be found the seed-states of all things....

寂静是具有创造性的土壤，一切事物的萌芽状态都能在这里找到。

——大卫·斯潘格勒

（DAVID SPANGLER）

Mon. Tue. Wed. Thur. Fri. Sat. Sun.

Date 年 月 日

Serenity is active. It is a gentle and firm participation with trust. Serenity is the relaxation of our cells into who we are and a quiet celebration of that relaxation.

宁静是活跃的，它是一种基于信任温和而坚定的参与。宁静让我们放松到本来的状态，并默默地庆祝这种放松。

——阿内·威尔逊·施费

（ANNE WILSON SCHAEF）

Tonight, let us honor silence.... May we respect the need for silence in our lives.

今夜，让我们向沉默致敬……愿我们尊重生活中沉默的必要性。

——M. J. 赖安

（M.J. RYAN）

Date

年

月

日

Tranquillity can often be the last, best gift of age.

平静往往是生命最后也是最好的礼物。

——帕姆·布朗

（PAM BROWN）

The poor long for riches and the rich for heaven, but the wise long for a state of tranquillity.

穷人渴望财富，富人渴望天堂，只有智者渴望平静。

——斯瓦米·拉玛
（SWAMI RAMA）

Mon. Tue. Wed. Thur. Fri. Sat. Sun.

Date

年

月

日

Until you make peace with who you are, you'll never be content with what you have.

除非你彻底接纳自己是谁，否则你永远无法满足于自己拥有的。

——多丽斯·莫特曼

（DORIS MORTMAN）

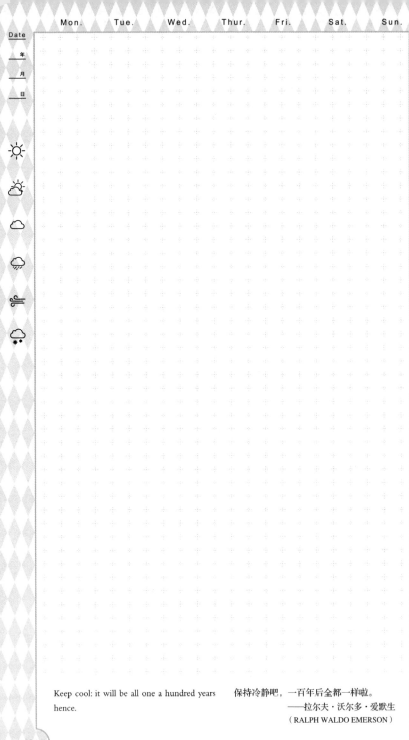

Keep cool: it will be all one a hundred years hence.

保持冷静吧，一百年后全都一样啦。

——拉尔夫·沃尔多·爱默生

（RALPH WALDO EMERSON）

Mon. Tue. Wed. Thur. Fri. Sat. Sun.

Date

年

月

日

Let us not look back in anger, nor forward in fear, but around us in awareness.

我们要不念过往，不惧将来，对身边保持觉察。

——詹姆斯·瑟伯

（JAMES THURBER）

My greatest wealth is the great stillness in which I strive and grow and win what the world cannot take from me with fire or sword.

我最大的财富便是沉静。我在其中奋斗成长，赢得世人无法用火与剑夺走的东西。

——约翰·沃尔夫冈·冯·歌德

（JOHANN WOLFGANG VON GOETHE）

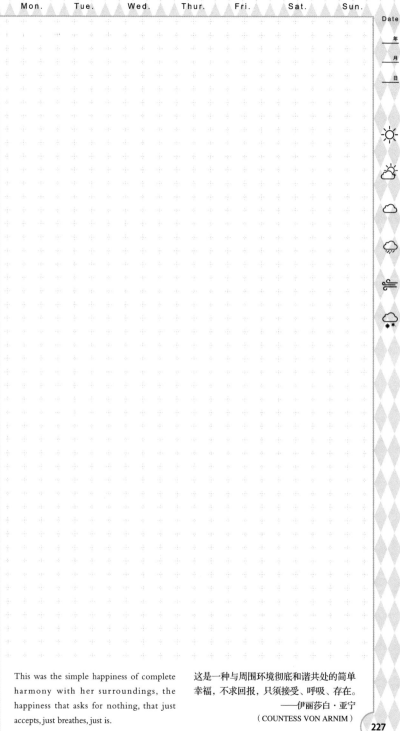

This was the simple happiness of complete harmony with her surroundings, the happiness that asks for nothing, that just accepts, just breathes, just is.

这是一种与周围环境彻底和谐共处的简单幸福，不求回报，只须接受、呼吸、存在。

——伊丽莎白·亚宁

（COUNTESS VON ARNIM）

The thing which we speak of as beauty does not have to be sought in distant lands.... It is here about us or it is nowhere....

我们所谓的美丽无须到遥远之处寻找……
它与我们同在，否则它便不存在……

——艾伦 · 塔克

（LLEN TUCKER）

Mon. Tue. Wed. Thur. Fri. Sat. Sun.

Date

年

月

日

I can only gaze at the universe	我只能凝视宇宙全部的真实形式,
In its full, true form,	凝望夜空中万点繁星,
At the millions of stars in the sky	承载无限和谐的美丽——
Carrying their huge harmonious beauty –	从不破坏节奏,
Never breaking their rhythm	从不走调,
Or losing their tune,	从不疯狂,
Never deranged	从不跌倒……
And never stumbling....	——罗宾德拉纳特·泰戈尔(ABINDRANATH TAGORE)

Over and over again I have experienced the quieting influence of rose scent upon a disturbed state of mind.

我一次又一次地体会到玫瑰花香能让心神不宁者平静安然。

——路易斯·毕比·怀尔德

（LOUISE BEEBE WILDER）

Flow with whatever may happen and let your mind be free. Stay centered by acceptance. This is the ultimate.

且夫乘物以游心，托不得已以养中，至矣。

——庄子

（CHUANG TZU）

I loaf and invite my soul.

我邀了我的灵魂同我一道闲游……

——沃尔特·惠特曼

（WALT WHITMAN）

Today. On a stone on his desk.

今日，刻于其放在书桌上的一块石头上

——约翰·拉斯金

（JOHN RUSKIN）

A quiet heart is a continual feast.

一颗平静心是一场持续不断的盛宴。

——谚语

（PROVERB）

Mon. Tue. Wed. Thur. Fri. Sat. Sun.

Date

年

月

日

Nature chose for a tool, not the earthquake or lightning to rend and split asunder, not the stormy torrent or eroding rain, but the tender snow flowers noiselessly falling through unnumbered centuries.

自然选择了一种工具，不是能让大地四分五裂的地震与闪电，不是湍急的洪流也不是侵蚀的雨水，而是柔软的雪花无声地飘落于无数个世纪。

——约翰·缪尔（JOHN MUIR）

Why is it that I find it so hard to take time for myself? Time to be, rather than time to do. And often what is urgent elbows its way to the forefront of my day and the important gets trampled in the rush.

为什么我发现很难有自己的时间？生活的时间，而不是忙碌的时间。紧急的事常常挤到一天的前头，重要的事则被匆匆践踏而过。

——佚名（AUTHOR UNKNOWN）

Mon. Tue. Wed. Thur. Fri. Sat. Sun.

Date

年

月

日

I hope you find joy in the great things of life – but also in the little things. A flower, a song, a butterfly on your hand.

我希望你既能在生活的大事中找到快乐，也能在小事中找到欢喜，比如一朵花、一首歌和手上的一蝴蝶。

——埃伦·列文

（ELLEN LEVINE）

At the centre of one's life one must retain a stillness – for this is where knowledge and experience are distilled to wisdom.

人必须在生活的中心保留一处平静之地——那里是知识和经验转化为智慧之处。

——帕姆·布朗

（PAM BROWN）

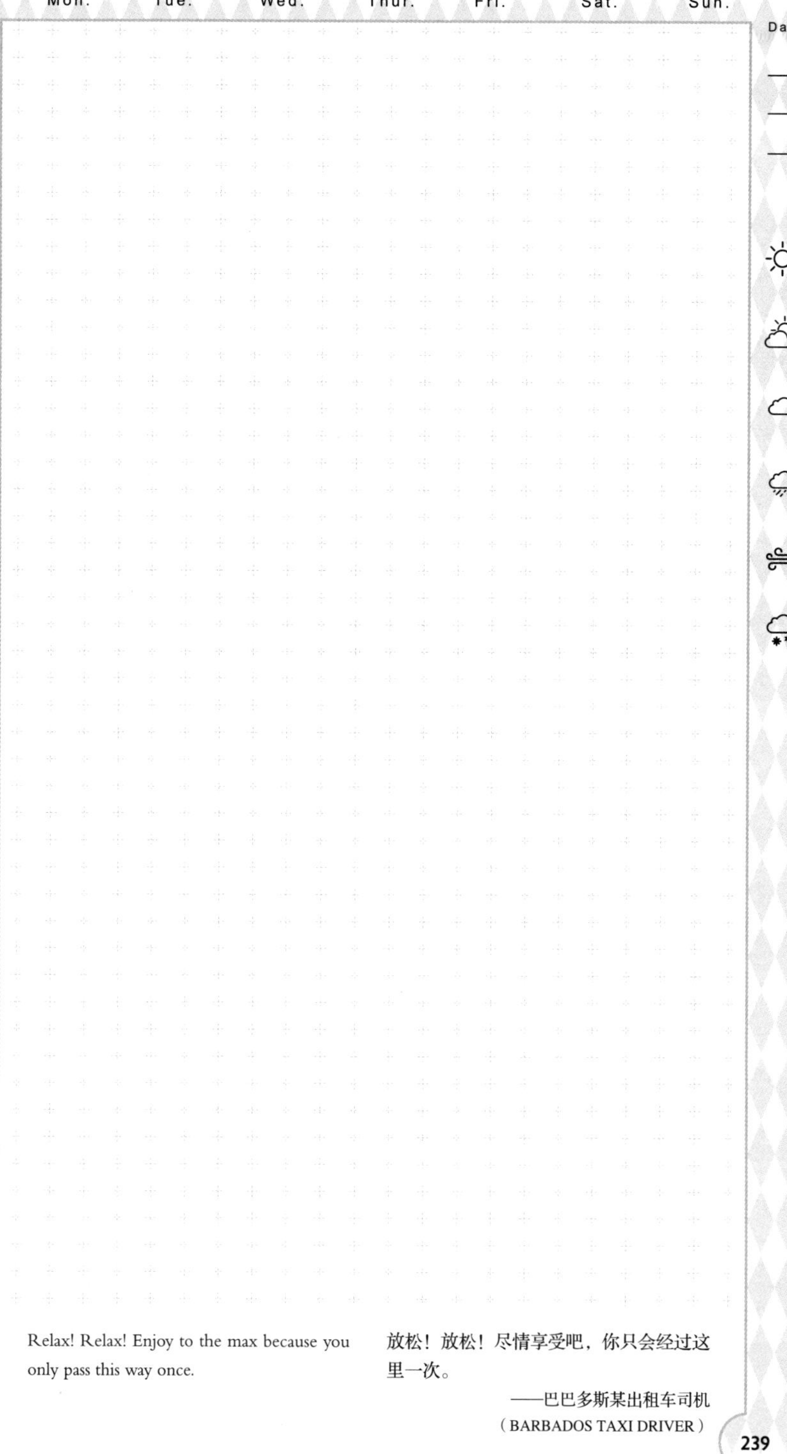

Relax! Relax! Enjoy to the max because you only pass this way once.

放松！放松！尽情享受吧，你只会经过这里一次。

——巴巴多斯某出租车司机

（BARBADOS TAXI DRIVER）

In the stones, the trees, and the skies, is a fulfilment for humanity, a contentment, without which no life can be satisfied or rested in the deepest sense.

在石头上、树林中和天空中都有着人类的成就感或满足感，没有它们，任何生命都无法满足，也无法获得最深刻的平静。

——鲍勃·布朗（BOB BROWN）

Be still. In small steps discard the clutter of possessions, the sound of thought. Let them go, till all that is left is quietness and peace.

保持平静。一小步一小步丢弃占有的杂念，以及来自思想的声音。放它们去吧，直到只剩下平静与安宁。

——帕姆·布朗
（PAM BROWN）

Mon. Tue. Wed. Thur. Fri. Sat. Sun.

Date

年

月

日

Solitude, quality solitude, is an assertion of self-worth, because only in the stillness can we hear the truth of our own unique voices.

孤独，高质量的孤独，是自我价值的一种肯定。因为只有在寂静中，我们才能听见自己独特声音的真相。

——珀尔·克利奇

（PEARL CLEAGE）

When we cannot find contentment in ourselves, it is useless to seek it elsewhere.

如果我们无法从自身得到满足，到别处寻找是没有用的。

——弗朗索瓦·德·拉罗什富科

（FRANÇOIS, DUC DE LA ROCHEFOUCAULD）

The air is full of sounds, sighs of the wind in the trees, sighs which fade back into the overhanging silence. A bee passes, a golden ripple in the quiet air.

空气中满是声音，有树木间风儿的叹息声，有消失在空悬寂静中的叹息声。一只蜜蜂飞过，安静的空气中荡漾出金色的涟漪。

——玛丽恩·米尔纳（MARION MILNER）

Each day provides its own gifts.

每一天都有自己的礼物。

——马提亚尔

（MARTIAL）

In character, in manner, in style, in all things, the supreme excellence is simplicity.

性格、态度、风格以及其他一切事物，简单的就是最好的。

——亨利·沃兹沃斯·朗费罗

（HENRY WADSWORTH LONGFELLOW）

Mon. Tue. Wed. Thur. Fri. Sat. Sun.

Date

年

月

日

Even if something is left undone, everyone must take time to sit still and watch the leaves turn.

即使有事没做完，每个人都应花时间静坐，看树叶飘落。

——伊丽莎白·劳伦斯

（ELIZABETH LAWRENCE）

I saw old Autumn in the misty morn stand shadowless like silence, listening to silence.

我在朦胧的晨曦中望见旧日的秋天寂静地站立，聆听寂静。

——托马斯·胡德（THOMAS HOOD）

All problems fade out in proportion as you develop this ability to be quiet, to behold and to witness divine harmony unfold.

当你能够安静下来，观看并见证神圣的和谐逐渐展开，所有的问题都会逐渐消失。

——乔尔·S. 戈德史密斯

（JOEL S. GOLDSMITH）

Only when we are silent can we begin to hear the voice that is truly our own – what the Quakers call "the still small voice within".

只有沉默的时候，我们才能听见自己真正的声音——贵格会信徒称之为“内心的微弱声音”。

——瑞克 · 菲尔德

（RICK FIELD）

There is only one way to happiness and that is to cease worrying about things which are beyond the power of our will.

快乐的方法只有一种，就是停止为我们无法控制的事担忧。

——爱比克泰德

（EPICTETUS）

I wish that life should not be cheap, but sacred. I wish the days to be as centuries, loaded, fragrant.

我希望生活不廉价，但充满神圣感。我希望每天都像几个世纪一样，收获满满，芬芳四溢。

——拉尔夫·沃尔多·爱默生
（RALPH WALDO EMERSON2）

Date

年

月

日

Ultimately we have just one moral duty: to reclaim large areas of peace in ourselves, more and more peace, and to reflect it toward others. And the more peace there is in us, the more peace there will be in our troubled world.

最终，我们只有一项道德责任：让自己的心灵获得广阔的平静，越来越多的平静，将其传递给他人。我们的心灵越平静，这骚动不安的世界也就越和平。

——艾蒂·西勒申

（ETTY HILLESUM）

Teach me the art of creating islands of stillness, in which I can absorb the beauty of everyday things: clouds, trees, a snatch of music....

请教我创造宁静之岛的艺术，让我能吸收日常事物之美：飘在空中的云朵，茂盛的树木，一段动听的音乐……

——玛丽安·斯特劳德

（MARION STROUD）

Mon. Tue. Wed. Thur. Fri. Sat. Sun.

Date

年

月

日

The mind is never right but when it is at peace within itself.

只有在内心平静的时候，头脑中的想法才是正确的。

——鲁齐乌斯·安奈乌斯·塞内加

（LUCIUS ANNAEUS SENECA）

Mon. Tue. Wed. Thur. Fri. Sat. Sun.

Date

年

月

日

If I am at peace with myself, it has been a successful day.

只要我内心平静，这就是成功的一天。

— 艾利克斯·诺布尔

（ALEX NOBLE）

Mon. Tue. Wed. Thur. Fri. Sat. Sun.

Date

年

月

日

Learn to be quiet enough to hear the sound of the genuine within yourself so that you can hear it in others.

学会安静到能听见自己内心真实的声音，那样你也能听见他人的。

——玛丽安·赖特·埃德尔曼
（MARIAN WRIGHT EDELMAN）

One flower at a time, I want to hear what it is saying.

一次一朵，我想听听花儿在说什么。

——罗伯特·弗朗西斯

（ROBERT FRANCIS）

Every day it's nice to stop and say, "Wait a minute! I am so lucky! This is great!"

每天停下来说几句“等一下！我太幸运了！真是太棒了！”这会很不错。

——凯特·哈德森

（KATE HUDSON）

Mon. Tue. Wed. Thur. Fri. Sat. Sun.

Date

年

月

日

Happiness is as a butterfly, which, when pursued, is always beyond our grasp, but which, if you will sit down quietly, may alight upon you.

快乐就像一只蝴蝶，要是主动去追，就总抓不到。但当你安静地坐下来，它就会降落在你手中。

——纳撒尼尔·霍桑

（NATHANIEL HAWTHORNE）

Mon.　Tue.　Wed.　Thur.　Fri.　Sat.　Sun.

Date

年

月

日

Mother Nature is kind to us. Live with nature. Go with the flow... live every minute NOW.

大自然母亲对我们十分友善。生于自然，顺其自然……活在当下。

——月羽公主奶奶

（GRANDMOTHER PRINCESS MOON FEATHERS）

A shaft of sunlight at the end of a dark afternoon, a note in music, and the way the back of a baby's neck smells.... Those are the important things.

黄昏结束前的最后 束阳光，一首乐曲的旋律，一个孩子脖子后的汗味儿……那些都是很重要的东西。

——E.B. 怀特
（E.B. WHITE）

Mon. Tue. Wed. Thur. Fri. Sat. Sun.

Date

年

月

日

Live in each season as it passes: Breathe the air, drink the drink, taste the fruit, and resign yourself to the influences of each.

好好度过每一个季节，呼吸新鲜的空气，畅饮美酒，品尝水果，尽情享受这一切。

——亨利·戴维·梭罗

（HENRY DAVID THOREAU）

Whenever you share love with another, you'll notice the peace that comes to you and to them.

每当与他人分享爱心，你都能注意到你和他们发自内心的平静祥和。

——特蕾莎修女

（MOTHER TERESA）

Mon. Tue. Wed. Thur. Fri. Sat. Sun.

Date

年

月

日

My life has no purpose, no direction, no aim, no meaning and yet I'm happy. I can't figure it out. What am I doing right?

我的生活没有目标，没有方向，没有意义，但我却很快乐。我不明白，我做得对吗？

——查尔斯·M. 舒尔茨

（CHARLES M. SCHULZ）

What lies behind us, and what lies before us are tiny matters, compared to what lies within us.

所有将要直面的，和已成为过往的东西，与埋在我们内心的，都是微末。

——拉尔夫·沃尔多·爱默生

（RALPH WALDO EMERSON）

Mon. Tue. Wed. Thur. Fri. Sat. Sun.

Date

年

月

日

Joy is the realization of oneness of our soul, the oneness of our soul with the world....

快乐就是我们灵魂合一的完成，同时灵魂与世界合一……

——罗宾德拉纳特·泰戈尔（RABINDRANATH TAGORE）

If you are losing your leisure, look out! You are losing your soul.

要是你正失去悠闲状态，当心！你正失去你的灵魂。

——罗根·皮尔萨尔·史密斯

（LOGAN PEARSALL SMITH）

To be interested in the changing seasons is a happier state of mind than to be hopelessly in love with spring.

充满兴致地感受四季的变迁，比无望地留恋春天要更加快乐。

——乔治·桑塔耶拿

（GEORGE SANTAYANA）

Those who know the value and the exquisite taste of solitary freedom (for one is only free when alone)...

那些懂得孤独的价值和高雅品味的人才自由（因为只有孤身一人时，人才真正自由）……

——伊莎贝尔·埃伯哈特
（ISABELLE EBERHARDT）

Mon. Tue. Wed. Thur. Fri. Sat. Sun.

Date

年

月

日

Teach me delight in simple things
And mirth that has no bitter springs;
Forgiveness free of evil done,
And love to all men' neath the sun.

教我为简单之物快乐，
没有苦泉的喜乐；
用宽恕摆脱罪恶，
爱天下人毫不吝啬。

——鲁德亚德 · 吉卜林
（RUDYARD KIPLING）

Learn to take a little time – even if it is a moment in the garden, a gallery, a café. Appreciate it. Let birds and frogs and pictures, music and books and undemanding friends restore you.

学会留一点时间——即使是在花园、画廊、咖啡馆里，尽情地享受它。让鸟儿、青蛙、图画、音乐、图书，以及随和的朋友助你得到恢复。

——帕姆·布朗（PAM BROWN）

Mon. Tue. Wed. Thur. Fri. Sat. Sun.

Date
年
月
日

A poor life this if, full of care, we have no time to stand and stare.

可怜的生活总是充满忧虑，让我们都没有时间伫立凝视。

——威廉·亨利·戴维斯
（WILLIAM HENRY DAVIES）

There is a luxury in being quiet in the heart of chaos.

在混乱的中心保持平静，是一种奢侈。

——弗吉尼亚·伍尔芙

（VIRGINIA WOOLF）

The way to use life is to do nothing through acting. The way to use life is to do everything through being.

道常无为，而无不为。

——老子
（LAO –TZU）

Mon. Tue. Wed. Thur. Fri. Sat. Sun.

Date

年

月

日

In addition to souls which run and shriek and devour might there not be souls which bloom in stillness....

除了那些奔跑、尖叫和被吞噬的，难道就没有在寂静中绽放的灵魂吗……

——彼得·汤普金斯和克里斯托弗·伯德

（PETER TOMPKINS AND CHRISTOPHER BIRD）

Date
年
月
日

Do not linger to gather flowers to keep them, but walk on, for flowers will keep themselves blooming all your way.

别为了采摘花朵停下脚步。继续前行，你的一路都会有花香相随。

——罗宾德拉纳特·泰戈尔

（RABINDRANATH TAGORE）

Mon. Tue. Wed. Thur. Fri. Sat. Sun.

Date

年

月

日

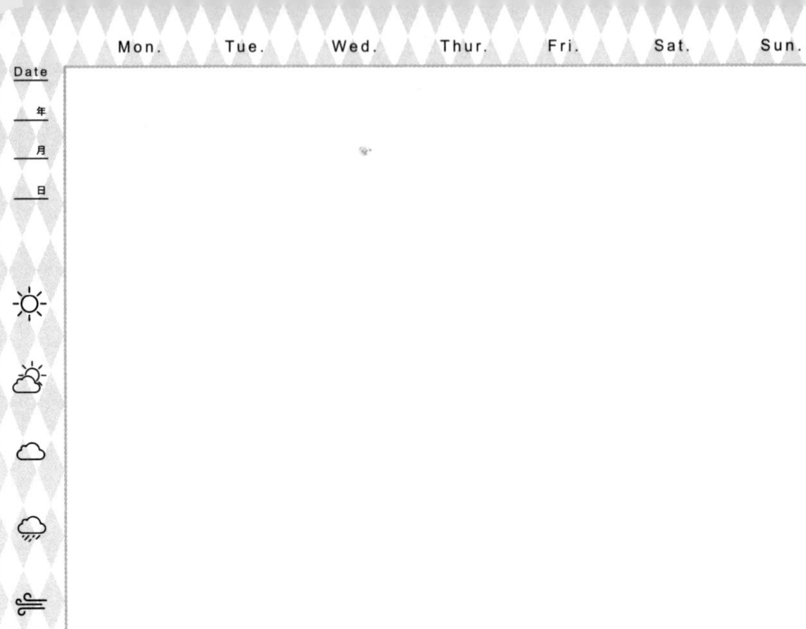

Time is really the only capital that any human being has and the only thing we can't afford to lose.

时间确实是人类唯一的资本，也是我们唯一丢不起的东西。

——托马斯·爱迪生

（THOMAS EDISON）

Date

年

月

日

Do what you can, with what you have, where you are.

在你所处的位置，用你所拥有的资源，做你所能做的事情。

——西奥多·罗斯福

（THEODORE ROOSEVELT）

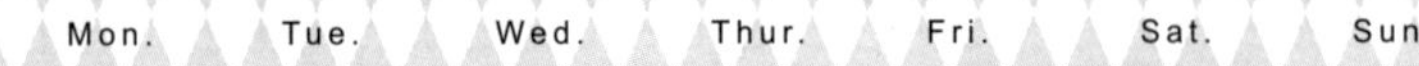

Date

年

月

日

Peace is not a passive but an active condition, not a negation but an affirmation. It is a gesture as strong as war.

和平不是被动而是主动状态，不是否定而是肯定。这种姿态与战争同样强大。

——玛丽·罗伯茨·莱因哈特

（MARY ROBERTS RINEHART）

Mon. Tue. Wed. Thur. Fri. Sat. Sun.

Date

年

月

日

Those who are firm, enduring, simple, and unpretentious are the nearest to virtue.

子曰："刚、毅、木、讷近仁。"

——《论语》

（THE ANALECTS）

Mon. Tue. Wed. Thur. Fri. Sat. Sun.

Date

年

月

日

The way you look at others, your smile, and your small acts of caring can create happiness. True happiness does not depend on wealth or fame.

你看待他人的方式、真诚的微笑、微小的善举都能带来快乐。真正的快乐不在于财富或名望。

——一行禅师

（THICH NHAT HANH）

Date

年

月

日

To preserve the silence within – amid all the noise. To remain open and quiet… no matter how many tramp across the parade-ground in whirling dust under an arid sky.

保持内心平静——即使身处噪音之中。保持开放与安宁……无论多少人在荒芜的天空下，纷飞的尘土间，踏过阅兵场。

——达格·哈马舍尔德

（DAG HAMMARSKJÖLD）

There is a time to keep silence and a time to speak.

静默有时，言语有时。

——《传道书》

（ECCLESIASTES）

Mon. Tue. Wed. Thur. Fri. Sat. Sun.

Date

年

月

日

When I am, as it were, completely myself, entirely alone, and of good cheer... it is on such occasions that ideas flow best and most abundantly.

我完全独自一人时，心情就十分愉悦……这才是灵感流动得最顺畅充足的时候。

——沃尔夫冈·阿玛多伊斯·莫扎特

（WOLFGANG AMADEUS MOZART）

The one who smiles rather than rages is always the stronger.

微笑者永远比愤怒者更强大。

——日本智慧

（JAPANESE WISDOM）

Mon. Tue. Wed. Thur. Fri. Sat. Sun.

Date
年
月
日

The present – the present is, after all, the only thing certain.

现在——毕竟，现在是唯一确定的事物。

——黛娜·玛丽亚·马洛克·克雷克
（DINAH MARIA MULOCK CRAIK）

It is good to be alone in a garden at dawn or dark so that all its shy presence may haunt you and possess you in a reverie of suspended thought.

黎明或黄昏时分，孤身一人待在花园挺好的。那样花园里所有羞涩的存在都将跑出来萦绕你，让你暂缓思考，陷入遐想。

——詹姆斯·道格拉斯（JAMES DOUGLAS）

Mon. Tue. Wed. Thur. Fri. Sat. Sun.

Date

年

月

日

I used to believe that anything was better than nothing. Now I know that sometimes nothing is better.

过去我曾相信有总比无好，现在我明白有些时候无比有强。

——格兰达·杰克逊

（GLENDA JACKSON）

Mon. Tue. Wed. Thur. Fri. Sat. Sun.

Date

年

月

日

It is good to have an end to journey toward; but it is the journey that matters, in the end.

旅途有个终点总是好的，但最重要的是旅途本身。

——厄修拉·勒古恩

（URSULA K. LE GUIN）

Life is eating us up. We shall be fables presently. Keep cool: it will be all one a hundred years hence.

生活正吞噬我们，我们将很快成为寓言。保持冷静吧，一百年后全都一样啦。

——拉尔夫·沃尔多·爱默生

（RALPH WALDO EMERSON）

Date

年

月

日

There's a crack, a crack in everything.
That's now the light gets in.

万物皆有裂痕，那是光照进来的地方。

——莱昂纳德·科恩

（LEONARD COHEN）

Enjoy the blessings of the day... and the evils bear patiently; for this day only is ours: we are dead to yesterday, and not born to tomorrow.

享受今日的祝愿吧……耐心地忍受罪恶。因为只有今日属于我们：对于昨日而言我们早已逝去，对于明日而言我们并未出生。

——杰瑞米·泰勒主教

（BISHOP JEREMY TAYLOR）

Mon. Tue. Wed. Thur. Fri. Sat. Sun.

Date

年

月

日

What life can compare to this? Sitting quietly by the window, I watch the leaves fall and the flowers bloom, as the seasons come and go.

争如独坐虚窗下，叶落花开自有时。

——释重显

（HSUEH–TOU）

Mon. Tue. Wed. Thur. Fri. Sat. Sun.

Date

年

月

日

We know nothing of tomorrow; our business is to be good and happy today.

我们对明天一无所知，重要的是顺利快乐地度过今天。

——悉尼·史密斯

（SYDNEY SMITH）

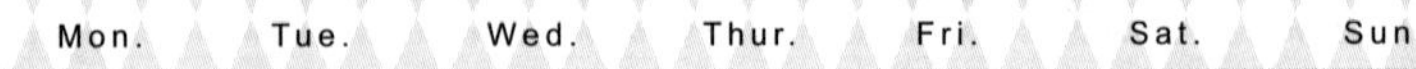

Date

年

月

日

I began to have an idea of my life, not as the slow shaping achievement to fit my preconceived purposes, but as the gradual discovery and growth of a purpose which I did not know.

我开始对我的生活有了概念，不是为达到预想目标而缓慢塑造成就，而是逐渐发现和培养我所不知道的目标。

——乔安娜·菲尔德

（JOANNA FIELD）

Mon. Tue. Wed. Thur. Fri. Sat. Sun.

Date

年

月

日

Too many people, too many demands, too much to do; competent, busy, hurrying people – It just isn't living at all.

太多的人，太多要求，太多要做的事。能干的、忙碌的、匆匆的人们——这根本不是生活。

——安妮·默洛·林德伯格

（ANNE MORROW LINDBERGH）

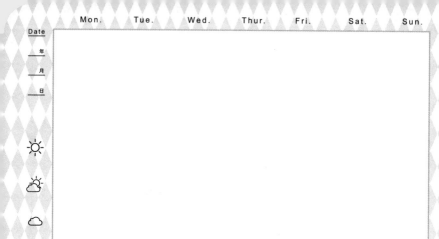

Knowing what you know, be serene also, like a mountain; and do not be distressed by misfortune.

知道你所知道的，也要像山一样宁静，不要为不幸而苦恼。

——萨纳伊

（SANAI）

Mon. Tue. Wed. Thur. Fri. Sat. Sun.

Date

年

月

日

We do not want riches. We want peace and love.

我们不要财富，我们要和平与友爱。

——红云

（RED CLOUD）

Mon. Tue. Wed. Thur. Fri. Sat. Sun.

Date

年

月

日

Cultivate solitude and quiet and a few sincere friends, rather than a mob of merriment, noise and thousands of nodding acquaintances.

交几个孤独、安静、真诚的朋友，而不是千万嘈杂、仅仅嬉闹的点头之交。

——威廉·鲍威尔

（WILLIAM POWELL）

Date

年

月

日

Deep in the soul, below pain, below all the distraction of life, is a silence vast and grand – an infinite ocean of calm, which nothing can disturb. Nature's own exceeding peace, which "passes understanding".

在心灵深处，生活一切纷扰与痛苦下，是一片宏伟辽阔的沉寂——一片无尽的平静之海，没有什么可以扰乱。那是自然本身的超然平静，"常人无法理解"的所在。

——R.M. 比克（R.M. BUCKE）

Mon. Tue. Wed. Thur. Fri. Sat. Sun.

Date

年

月

日

It is eternity now. I am in the midst of it. It is about me in the sunshine; I am in it, as the butterfly floats in the light-laden air. Nothing has to come; it is now.

此刻就是永恒，我正身处其中。我置身阳光之下，就像蝴蝶在轻盈的空气中飞舞。什么都无须到来，就是当下的一切。

——理查德·杰弗里斯

（RICHARD JEFFERIES）

Date
年
月
日

What is more tranquil than a musk-rose blowing in a green island far from all men's knowing?

什么比麝香玫瑰更为安隐，开在翠岛上远离世人记忆？

——约翰·济慈
（JOHN KEATS）

Mon. Tue. Wed. Thur. Fri. Sat. Sun.

Date

年

月

日

All the truly deep people have at the core of their being the genius to be simple or to know how to seek simplicity.

所有真正深刻的人内心都有一种天赋，懂得要简单或如何追求简单。

——马丁·埃米尔·马蒂（MARTIN MARTY）

Date

年

月

日

It is when you are really living in the present – working, thinking, lost, absorbed in something you care about very much, that you are living spiritually.

只有真正活在当下——努力工作，认真思考，陷入迷思，专注于你所关心的事物，你才有精神生活。

——布兰达·尤兰

（BRENDA UELAND）

Serenity can be at the heart of the busiest person. It is the quietness at the core of being.

最忙碌的人内心也能有宁静，这是居于存在之中心的安宁。

——帕姆·布朗

（PAM BROWN）

Mon. Tue. Wed. Thur. Fri. Sat. Sun.

Date

年

月

日

It is difficult to live in the present, ridiculous to live in the future and impossible to live in the past. Nothing is as far away as one minute ago.

活在当下是困难的，活在未来是荒谬的，活在过去是不可能的。没有什么比一分钟前更遥远。

——吉姆 · 毕肖普

（JIM BISHOP）

We cannot long survive without air, water, and sleep. Next in importance comes food. And close on its heels, solitude.

没有空气、水和睡眠，我们无法长久生存。其次是食物。紧随其后的便是孤独。

——托马斯·萨斯

（THOMAS SZASZ）

Mon. Tue. Wed. Thur. Fri. Sat. Sun.

Date

年

月

日

We should be blessed if we took advantage of every accident that befell us, like the grass which confesses the influence of the slightest dew that falls on it....

我们要是好好利用发生在我们身上的每件事情，就会获得祝福，就像草叶感激落于其上的每滴露珠对它的恩惠……

——亨利·戴维·梭罗
（HENRY DAVID THOREAU）

Mon. Tue. Wed. Thur. Fri. Sat. Sun.

Date

年

月

日

Grant yourself a moment of peace and you will understand how foolishly you have scurried about. Learn to be silent and you will notice that you have talked too much.

气收自觉怒平，神敛自觉言简。容人自觉味和，守静自觉天宁。

——陈继儒

（TSCHEN TSCHI JU）

Mon. Tue. Wed. Thur. Fri. Sat. Sun.

Date

年

月

日

Step out of the interesting, dynamic rhythm every so often and focus on your internal life. Say "Stop the world, I want to get off" for a while at least.

时不时走出有趣而充满活力的节奏，专注于你的内心生活。过段时间至少说一次"世界快停下，我要下车"。

——内奥米·罗森布莱特（NAOMI ROSENBLATT）

To see the world in a grain of sand
And Heaven in a wild flower....

一沙一世界，一花一天堂。

——威廉·布莱克
（WILLIAM BLAKE）

Mon. Tue. Wed. Thur. Fri. Sat. Sun.

Date

年

月

日

No land belongs unto me, yet I can go out in the meadows and see the healthy green grass – and behold the shower fall, and he that feels this, who can say he is poor?

我没有一片土地，却能走到草地上，看绿草健康生长，见阵雨落下。能感受到这一切的人，谁能说他贫穷呢？

——约翰·克莱尔

（JOHN CLARE）

Yesterday is a cancelled cheque; tomorrow is a promissory note; today is the only cash you have – so spend it wisely.

昨日是作废的支票，明日是约定的期票，只有今日是手中的现金——请物尽其用。

——凯·莱昂斯
（KAY LYONS）

Every form of happiness is private. Our greatest moments are personal, self-motivated, not to be touched.

任何形式的幸福都是私人的。我们最伟大的时刻都属于个人，是自觉自愿的，不容触碰。

——艾茵·兰德

（AYN RAND）

Mon. Tue. Wed. Thur. Fri. Sat. Sun.

Date

年

月

日

Every day, stop before something beautiful long enough to say, "Isn't that beautiful!"

每一大，都要在美丽的事物前长久驻足，说："多美啊！"

——艾莉丝·弗里曼·帕默

（ALICE FREEMAN PALMER）

Mon. Tue. Wed. Thur. Fri. Sat. Sun.

Date

年

月

日

Will is of little importance, complaining is nothing, fame is nothing. Openness, patience, receptivity, solitude is everything.

意愿无关紧要，抱怨与名望毫无用处。保持开放、耐心、接纳的心态，孤独就是一切。

——赖内·马利亚·里尔克
（RAINER MARIA RILKE）

Mon. Tue. Wed. Thur. Fri. Sat. Sun.

Date

年

月

日

To live content with small means; to seek elegance rather than luxury, and refinement rather than fashion; to be worthy, not respectable, and wealthy, not rich; to study hard, think quietly, talk gently, act frankly....

收入不高，但要活得知足；追求优雅，而不是奢华；追求教养，而非时尚；要有价值，但不必被社会认可；要富足，但不用特别富有。努力学习，安静思考，温柔交谈，坦诚待人……

——威廉·埃勒里·钱宁（WILLIAM ELLERY CHANNING）

Date

年

月

日

You will find that deep place of silence right in your room, your garden or even your bathtub.

你会在你的房间、花园甚至浴缸里，找到那深沉的宁静。

——伊丽莎白·库伯勒·罗斯

（ELISABETH KÜBLER–ROSS）

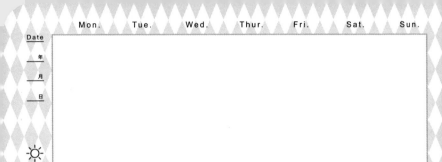

In the ancient world it was ever the greatest of the emperors and the wisest of the philosophers that sought peace and rest in a garden.

古代世界，最伟大的帝皇和最明智的哲学家都曾在花园中休憩，寻求平静。

——乔治·西特韦尔爵士

（SIR GEORGE SITWELL）

Fools are aye fond o' flittin', and wise men o' sittin'.

愚蠢的人喜欢走动，聪明的人喜欢静坐。

——苏格兰谚语

（SCOTTISH PROVERB）

Mon. Tue. Wed. Thur. Fri. Sat. Sun.

Date

年

月

日

It seems to me that one of the greatest stumbling blocks in life is this constant struggle to reach, to achieve, to acquire.

在我看来，生命中的最大绊脚石是不断奋力去追逐目标、获得成功、取得成就。

——吉杜·克里希那穆提

（JIDDU KRISHNAMURTI）

Better than a thousand useless words is one single word that gives peace.

带来宁静的只言片语，胜过万千无用之言。
——《达马帕达》
（THE DHAMMAPADA）

Mon. Tue. Wed. Thur. Fri. Sat. Sun.

Date

年

月

日

Our language has wisely sensed the two sides of being alone. It has created the word "loneliness" to express the pain of being alone. And it has created the word "solitude" to express the glory of being alone.

我们的语言睿智地捕捉到了“独处”的两个方面。“孤独”用于表达独处的痛苦，“清静”用于表达独处的美好。

——保罗·田立克

（PAUL TILLICH）

Between the in-breath and the out-breath lies the possibility of the future.

呼吸之间，未来就有无限可能。

——瑞沙德·菲尔德
（RESHAD FIELD）

Mon. Tue. Wed. Thur. Fri. Sat. Sun.

Date

年

月

日

Such sweet content, such minds, such sleep, such bliss, Beggars enjoy when princes oft do miss.

如此甜美的满足，如此的思想，如此的睡眠，如此的极乐，乞儿乐在其中，王子常常错过。

——罗伯特·格林

（ROBERT GREENE）

Date
年
月
日

Learn the sweet magic of a cheerful face; Not always smiling, but at least serene.

了解令人愉悦的脸庞所拥有的奇妙魔力：不用一直微笑，但至少保持平静。

——奥利弗·温德尔·霍姆斯

（OLIVER WENDELL HOLMES）

Date
年
月
日

Like water which can clearly mirror the sky and the trees only so long as its surface is undisturbed, the mind can only reflect the true image of the self when it is tranquil and wholly relaxed.

水面只有在波澜不惊的时候，才能清晰地倒映蓝天绿树。同样地，思想只能在完全平静放松的状态下，才能反映真实的自我。

——英德拉·黛维
（INDRA DEVI）

Date
年
月
日

Without stirring abroad one can know the whole world; Without looking out of the window one can see the way of heaven. The further one goes the less one knows.

不出于户，以知天下；不窥于牖，以知天道。其出也弥远，其知弥少。

——老子
（LAO-TZU）

Mon. Tue. Wed. Thur. Fri. Sat. Sun.

The sacred is not in the sky, but rather is based on this earth, in the ordinary dwelling places of our lives, in our gardens and kitchens and bedrooms.

神圣的事物不在天上，而在地上，在随处可见的我们生活的房子里，在花园里，在厨房里，在卧室里。

——玛莉莲·休厄尔

（MARILYN SEWELL）

Mon. Tue. Wed. Thur. Fri. Sat. Sun.

Date

年

月

日

One cannot appreciate beauty on the run. When I can be motionless long enough, there is no limit I have ever reached to the revelations in an opening bud.

人在奔走的时候无法欣赏美。如果我可以在足够长的时间内静止不动，就算是一朵绽放的花蕾，也会给我无穷的启发。

——维达·达顿·斯卡德

（VIDA D. SCUDDER）

Mon. Tue. Wed. Thur. Fri. Sat. Sun.

Date

年

月

日

The people are blessed who every day are permitted to behold anything so pure and serene as the western sky at sunset, while revolutions vex the world.

在革命烦扰全世界之时，那些可以看到每天日落时西边天空这样纯洁宁静的景色的人，可真是太幸福了！

——亨利·戴维·梭罗

（HENRY DAVID THOREAU）

Date

年

月

日

Over all the mountaintops is peace. In all treetops you perceive scarcely a breath. The little birds in the forest are silent. Wait then; soon You, too, will have peace.

所有山巅之上，是平静。所有树梢之上，是几乎听不到的呼吸声。森林中的小鸟，一片静默。等待，很快，你也将拥有平静。

——约翰·沃尔夫冈·冯·歌德

（JOHANN WOLFGANG VON GOETHE）

Mon. Tue. Wed. Thur. Fri. Sat. Sun.

Date

年

月

日

Speak nought, move not, but listen, the sky is full of gold. No ripple on the river, no stir in field or fold, All gleams but nought doth glisten, but the far-off unseen sea.

不要说话，不要走动，静静聆听，天空金黄一片。河面上没有一丝涟漪，田地里、围栏里没有任何动静。除了远处望不到头的河流，所有的东西都闪着微光，却并不刺眼。

——威廉·莫里斯（WILLIAM MORRIS）

Mon. Tue. Wed. Thur. Fri. Sat. Sun.

Date

年

月

日

This true simplicity makes us conscious of a certain openness, gentleness, innocence, gaiety and serenity.

这种至真的简单让我们感受到了某种开放、温柔、天真、愉悦和宁静。

——弗朗索瓦·芬乃伦

（FRANÇOIS FÉNELON）

Mon.　Tue.　Wed.　Thur.　Fri.　Sat.　Sun.

Date

年

月

日

In quietness one comes to know that nothing good is ever lost. It has become a part of all that is, and all that is to come.

在平静中，我们意识到：美好的事物从不曾消失。它们只是融入了所有已存在的事物，以及将要出现的事物中。

——帕姆·布朗

（PAM BROWN）

Mon. Tue. Wed. Thur. Fri. Sat. Sun.

Date

年

月

日

One cannot collect all the beautiful shells on the beach. One can collect only a few, and they are more beautiful if they are few.

谁也无法将沙滩上所有漂亮贝壳捡起收藏。他只能捡起其中几只，如果这些贝壳属于稀少类型，那就显得更加美丽。

——安妮·默洛·林德伯格

（ANNE MORROW LINDBERGH）

Mon. Tue. Wed. Thur. Fri. Sat. Sun.

Date

年

月

日

The strong, calm person is always loved and revered. That person is like a shade-giving tree in a thirsty land, or a sheltering rock in a storm.

坚强冷静的人总是受人爱戴和尊敬。他们像是干旱之地的遮荫树，又如暴风雨中的磐石，为人遮阳挡雨。

——詹姆斯·艾伦

（JAMES ALLEN）

Date
年
月
日

The desert silences, the murmuring oceans, the wind-brushed moors, the mountain tops, all held in the centre of your being. Here is the stillness that you seek.

寂静的沙漠，低吟的海洋，风声簌簌的荒原，巍峨的山巅，都在你的生命中心。这就是你所寻求的宁静。

——帕姆·布朗
（PAM BROWN）

Mon. Tue. Wed. Thur. Fri. Sat. Sun.

Date

年

月

日

It is important to stay close enough to the pulse of life to feel its rhythm, to be comforted by its steadiness, to know that life is vital, and one's own minute living a torn fragment of the larger cloth.

贴近生活的脉搏，感受它的节奏，在平淡中治愈心灵。了解生命的重要性，知道每个人渺小的生命都是宏大布景中扯下来的一小块，这些都是很重要的。

——玛·金·罗琳斯

（MARJORIE KINNAN RAWLINGS）

Mon. Tue. Wed. Thur. Fri. Sat. Sun.

Date

年

月

日

Slow down and enjoy life. It's not only the scenery you miss by going too fast – you also miss the sense of where you're going and why.

放慢脚步，享受生活。倘若步履匆匆，你不仅会错过一路风景，还无暇思考你去向何处，为何为出发。

——埃迪·康托尔
（EDDIE CANTOR）

There is no music in a "rest" but there is the making of music in it.

当音乐奏到休止符时，声音完全停止。这常常是音乐最精彩之处。

——约翰·拉斯金

（JOHN RUSKIN）

Mon. Tue. Wed. Thur. Fri. Sat. Sun.

Date

年

月

日

Half of the confusion in the world comes from not knowing how little we need.... I live more simply now, and with more peace.

世界上一半的困惑来源于不知道自己需要的其实很少……现在我的生活更简单，获得了更多的宁静。

——李察·波德

（RICHARD EVELYN BYRD）

Mon. Tue. Wed. Thur. Fri. Sat. Sun.

Date

年

月

日

Be at peace and see a clear pattern running through all your life.

保持平静，就能清晰地省察到你的人生模式。

——艾琳·卡迪

（EILEEN CADDY）

Mon. Tue. Wed. Thur. Fri. Sat. Sun.

Date

年

月

日

Not real stillness, but just the trees,
Low whispering, or the hum of bees,
Or brooks faint babbling over stones,
In strangely, softly tangled tones....

不是真正的宁静，只是树木，
喃喃低语，或蜜蜂嗡嗡作响，
或是小溪流过石块发出潺潺的声音。
所有的声音，以奇怪而轻柔的方式，缠结在一起……

——尼克松·沃特曼（NIXON WATERMAN）

Date

年

月

日

Make friends with the dark and silence. Make them a refuge from the tumult of the world.

与黑暗和寂静做朋友，让它们成为你远离世上纷扰的庇护所。

——帕姆·布朗

（PAM BROWN）

Mon. Tue. Wed. Thur. Fri. Sat. Sun.

Date

年

月

日

All real and wholesome enjoyments possible to us have been just as possible to us, since first we were made of the earth, as they are now: and they are possible to us chiefly in peace.

从人类诞生开始，到现在为止，我们一直都有可能获得一切真正的、有益的快乐。但要得到这些快乐，首先要有平和的环境。

——约翰·拉斯金（JOHN RUSKIN）

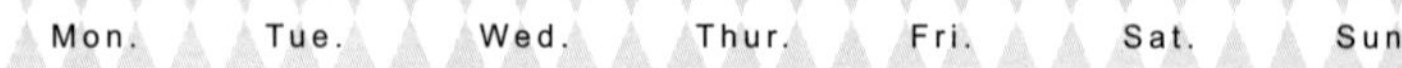

Date

年

月

日

Calm when the world is full of sound and fury. Love when the world seems cruel and heartless....My wish for you.

当世界充满喧嚣与狂乱时，保持镇静；当世界残酷无情时，怀有友爱之心。这就是我对你们的祝福。

——帕姆·布朗

（PAM BROWN）

Mon. Tue. Wed. Thur. Fri. Sat. Sun.

Date

年

月

日

Time and again we miss out on the great treasures in our lives because we are so restless. In our minds we are always elsewhere. We are seldom in the place where we stand and in the time that is now.

我们一次又一次地错过了生活中的宝贵财富，因为我们是如此的焦躁不安。在我们的意识里，我们总是身在别处。我们很少把注意力放在脚下的土地和当下的时间上。

——约翰·欧多诺休（JOHN O’ DONOHUE）

Mon. Tue. Wed. Thur. Fri. Sat. Sun.

Date

年

月

日

Work is not always required of us. There is such a thing as sacred idleness, the cultivation of which is now fearfully neglected.

我们不一定要工作。闲散看起来是非常神圣的，但我们却忽略了养成闲散的习惯，这实在可怕。

——乔治·麦克唐纳

（GEORGE MACDONALD）

Date
年
月
日

Wanting is the urge for the next moment to contain what this moment does not. When there's wanting in the mind, the moment feels incomplete. Wanting is seeking elsewhere. Completeness is being right there.

欲望是迫切希望下一刻获得此刻没有的东西。如果头脑里有欲望，这一刻就显得不完整。欲望是去别处寻找，完整就是老老实实待在原地。

——史蒂芬·莱文（STEPHEN LEVINE）

There is great happiness in not wanting, in not being something, in not going somewhere.

清心寡欲，安于一隅是极大的快乐。

——吉杜·克里希那穆提

（JIDDU KRISHNAMURTI）

Mon. Tue. Wed. Thur. Fri. Sat. Sun.

Date

年

月

日

Sit, rest, work. Alone with yourself, never weary. On the edge of the forest. Live joyfully, without desire.

坐定，休息，劳作。独自一人，从不厌倦。在森林边，快乐地生活，无欲无求。

——释迦牟尼

（THE BUDDHA）

Mon. Tue. Wed. Thur. Fri. Sat. Sun.

Date

年

月

日

The serene have not opted out of life. They see more widely, love more dearly, rejoice in the things the frantic mind no longer sees or hears.

平静的人并没有脱离生活。他们的视野更宽广，爱恋更深刻。那些手忙脚乱的人看不到、听不到的东西，他们却能感受到并心生欣喜。

——帕姆·布朗（PAM BROWN）

Date

年

月

日

I also know that when I'm trusting and being myself as fully as possible, everything in my life reflects this, by falling into place easily, often miraculously.

我还知道，当我尽可能地信任他人，做我自己的时候，我生命中的一切都会反映出这一点：事情很容易有条不紊地进行，而且常常是奇迹般地呈现出来。

——莎克蒂·高文（SHAKTI GAWAIN）

Mon. Tue. Wed. Thur. Fri. Sat. Sun.

Date

年

月

日

When we are in the present moment, our work on earth begins.

当我们身处当下，我们的使命就开始了。

——瑞沙德·菲尔德

（RESHAD FIELD）

Mon. Tue. Wed. Thur. Fri. Sat. Sun.

Date

年

月

日

Happiness is wanting your past, present and future just the way it is, was and always will be. Not wishing for the other.

快乐是接纳过去、现在、未来应有的样子，而不是奢望其它。

——《大地舞鼓：生命的礼赞》）

（EARTH DANCE DRUM）

Mon. Tue. Wed. Thur. Fri. Sat. Sun.

Date

年

月

日

There is no way in which to understand the world without first detecting it through the radar-net of our senses.

要了解世界，必须首先通过我们感官，像雷达网进行探测。

——黛安娜·阿克曼

（DIANE ACKERMAN）

Mon. Tue. Wed. Thur. Fri. Sat. Sun.

Date

年

月

日

Take time to dream – It is hitching your wagon to a star. Take time to love and to be loved – Take time to look around – It is too short a day to be closed in. Take time to laugh – It is the music of the soul.

花点时间去追求梦想，那说明你树立雄心壮志；花点时间去爱与被爱；花点时间到处看看，毕竟一天转瞬即逝；花点时间放声大笑，那是心灵的音乐。

——古英语祈祷文（OLD ENGLISH PRAYER）

Mon. Tue. Wed. Thur. Fri. Sat. Sun.

Date

年

月

日

Nobody sees a flower really; it is so small. We haven't time, and to see takes time – like to have a friend takes time.

没有人真正看到一朵花；它是如此渺小。我们没有时间，而观察花是需要时间的——就像交朋友需要时间一样。

——乔治亚·奥基弗

（GEORGIA O' KEEFFE）

Mon. Tue. Wed. Thur. Fri. Sat. Sun.

Date

年

月

日

The heart of the wise person lies quiet like limpid water.

智者的心灵平静如水。

——喀麦隆谚语

（CAMEROONIAN SAYING）

Mon. Tue. Wed. Thur. Fri. Sat. Sun.

Date

年

月

日

It becomes necessary to learn how to clear the mind of all clouds, to free it of all useless ballast and debris by dismissing the burden of too much concern with material things.

学习如何驱散头脑中的阴云至关重要，要卸下世俗烦恼的负担，扫开沉重的巨石和杂乱的残骸。

——英德拉·黛维

（INDRA DEVI）

Date 年 月 日

While others miserably pledge themselves to the insatiable pursuit of ambition and brief power, I will be stretched out in the shade singing.

当其他人可悲而无止境地滋生野心，追求稍纵即逝的权力时，我将舒展身体，在树荫下尽情歌唱。

——弗赖·路易斯·德·莱昂
（FRAY LUIS DE LEÓN）

That which we seek with passionate longing, here and there, upward and outward; we find at last within ourselves.

我们上下求索、四处寻觅的东西，到最后才发现，其实一直就在我们自己身上。

——理查德·莫里斯·比克（R.M.BUCKE）

Date
年
月
日

Life forms illogical patterns. It is haphazard and full of beauties which I try to catch as they fly by, for who knows whether any of them will ever return?

生活画出了毫无章法的图案。它充满偶然性，以及各种美的事物。我试图抓住这些即将逝去的美丽，因为没人知道它们还会不会再回来。

——玛戈特·芳廷（DAME MARGOT FONTEYN）

365 Calm Days: A peaceful thought for every day
Published in 2016 by Helen Exley Giftbooks in Great Britain.
Photography by Richard Exley © Helen Exley Greative Ltd 2016.
Selection and arrangement by Helen Exley © Helen Exley Greative Ltd 2016.

Printed in China.
www.helenexleygiftbooks.com

The Simplified Chinese translation rights are arranged through RR Donnelley Asia

图书在版编目（CIP）数据
微风吹拂：自在手账：英汉对照 /（英）海伦·埃克斯利编选；周成刚主编．-- 北京：新星出版社，2021.3
（新新悦读）
ISBN 978-7-5133-4230-8
Ⅰ．①微… Ⅱ．①海… ②周… Ⅲ．①本册②格言－汇编－世界－汉、英
Ⅳ．① TS951.5 ② H033
中国版本图书馆 CIP 数据核字（2020）第 249176 号

微风吹拂：自在手账

编　　选：[英] 海伦·埃克斯利
丛书主编：周成刚

策划编辑：李金学
责任编辑：姜　淮
特约编辑：赵　丹
责任校对：刘　义
责任印制：李珊珊
装帧设计：冷暖儿

出版发行：新星出版社
出 版 人：马汝军
社　　址：北京市西城区车公庄大街丙3号楼　100044
网　　址：www.newstarpress.com
电　　话：010-88310888
传　　真：010-65270449
法律顾问：北京市岳成律师事务所

读者服务：010-88310811　service@newstarpress.com
邮购地址：北京市西城区车公庄大街丙3号楼　100044

印　　刷：北京尚唐印刷包装有限公司
开　　本：889mm×1194mm　1/32
印　　张：11.75
字　　数：50千字
版　　次：2021年3月第一版　2021年3月第一次印刷
书　　号：ISBN 978-7-5133-4230-8
定　　价：68.00元